Heat and Mass Transfer

Series editors

D. Mewes, Hannover, Germany
F. Mayinger, Garching, Germany

For further volumes:
http://www.springer.com/series/4247

Achintya Kumar Pramanick

The Nature of Motive Force

Springer

Achintya Kumar Pramanick
Department of Mechanical Engineering
National Institute of Technology Durgapur
West Bengal
India

ISSN 1860-4846 ISSN 1860-4854 (electronic)
ISBN 978-3-642-54470-5 ISBN 978-3-642-54471-2 (eBook)
DOI 10.1007/978-3-642-54471-2
Springer Heidelberg New York Dordrecht London

Library of Congress Control Number: 2014932827

Printed on acid-free paper

Springer is part of Springer Science+Business Media (www.springer.com)

Decisively Dedicated
To
My Parents,
My Guru Srimat Swami Paramananda
and
My Better Half

"Perhaps this monsoon has poured more;
More than a sobbing soul before a mirage.
Bushes are hiding a secret love
Under the bosom of damsel earth.
Didn't you listen the nocturnal flute?

The lusty green grasses in my lawn
Smiling with unusual shyness.
The virgin hibiscus is still giggling
For she is untouched tonight.
Can't you smell the seducing fragrance?

The morning breeze is heavy today.
The temple songs are more melodious.
So I forgo to prepare my lecture notes:
The diary of a man's search.
Why don't you see the chariot is ready?"

Achintya Kumar Pramanick

When I heard the learn'd astronomer,

When the proofs, the figures, were ranged in columns before

me,

When I was shown the charts and diagrams, to add, divide,

and measure them,

When I sitting heard the astronomer where he lectured with

much applause in the lecture-room,

How soon unaccountable I became tired and sick,

Till rising and gliding out I wander'd off by myself,

In the mystical moist night-air, and from time to time,

Look'd up in perfect silence at the stars.

W. Whitman

Foreword

Professor Pramanick's "The Nature of Motive Force" is a delightful walk through the garden of thermodynamics and design in nature. For those who know thermodynamics, this book and its many ideas, quotes, and references are a treat. For those who are curious and eager to know, this is a very attractive invitation.

The law of motive force is the general observation that one or more tradeoffs happen when an effort is made to effect a change. From such tradeoffs emerge the features of organization that persist (shape, dimensions, structure, rhythm). This natural tendency is illustrated with numerous examples from thermal sciences: thermodynamics, heat transfer, and fluid mechanics.

"The Nature of Motive Force" is a treatise on the beauty and permanence of thermodynamics. It puts together several contemporary advances such as the constructal law, the intersection of asymptotes, entropy generation minimization, and convection fundamentals. The book derives its strength from Prof. Pramanick's erudition and strong grasp of mathematics, thermodynamics, languages, and history. I recommend this book very strongly.

Adrian Bejan
J. A. Jones Distinguished Professor
Duke University

Preface

'Tis the good reader that makes the good book; in every book he finds passages which seem confidences or asides hidden from all else and unmistakenly meant for his ear; the profit of books is according to the sensibility of the reader; the profoundest thought or passion sleeps as in a mine, until it is discovered by an equal mind and heart.

R. W. Emerson

I was always very curious about how one gets around to preparing a book and so perhaps as a reader you are as well. Mark Twain takes me into his confidence with the words: "...there ain't nothing more to write about, and I am rotten glad of it, because if I'd'a' knowed what trouble it was to make a book I wouldn't'a' tackled it, and ain't a-going to no more." In contemporary practice portraying a prelude many not have *argumentum ad hominem* much in vogue. But in the entire gamut of my reading experience, I never laid hands on any treatise without going through the very pursuit of the author first. On the same ground, it remains almost a compelling choice for me to insinuate what inspires me. I plead to exempt me from the *egalitarian fallacy* of trying to make all persons alike.

Every true research is but autobiographical and so is the following monograph. At a personal level, trying the best to be very meticulous and carping on almost every aspect that crops up in my way even results in imperfect performance, and thus further suffering a setback of dilemma on decision. Riding on the lacuna of my habit of witnessing ill decision and the stigma of a perfectionist, I was prompted to compose my first scientific writing [1] while I was a second-year undergraduate student in 1991, of a 4-year Mechanical Engineering degree program at the National Institute of Technology Durgapur, India, formerly recognized as Regional Engineering College Durgapur. After many years of latency, in January 2007, I submitted my doctoral thesis [2] haunted by my way of dogma and dilemma and by June 2007 I defended. In 2009, my doctoral thesis was selected only in the group of top five by the Prigogine prize selection committee for the best doctoral thesis in thermodynamics and hence my work could not see the delightful sun of scrutiny by a wide range of readers. Today, I continue to regard that my scientific approach has not been well circulated, especially among physicists. Until in January 2012, when I got a call from Adrian Bejan to publish a book chapter [3], I did not get a pat on my back. By now I got older and somewhat more immuned and case hardened about what other people would think of my preparation and presentation. Granted by Heaven, maybe I can afford to toy myself with the fascinating idea of writing a book. Jonathan Swift has rightly pointed out that, "if a Writer would know how to behave himself with relation to Posterity; let him consider in old Books, what he finds, that he is glad to know; and what Omissions he most laments."

The essay by Sadi Carnot [4], about a quarter of a century earlier than the terminology adopted by Thomson [5], is a milestone example of how the proponent of a new theory has no choice but to misuse the language of old theory [6]. Thus, without a constant misuse of language there cannot be any discovery, any progress [7]. Tuesdell [8, 9] addressed the celebrated failure of thermodynamics in the nineteenth century, accursed by misunderstanding, irrelevance, and retreat. In the spreaded span of the late twentieth century to the beginning of the twenty-first century, through the constructal theory (fourth law of thermodynamics) [10–12] proposed by Bejan, a consistent brilliant progress has been made in the unified description of nature as well as artificial (engineered) systems. Leib and Yngvason, [13] for the first time in the history of thermodynamics, made it scholarly possible to realize the concept of entropy purely on a macroscopic basis, in contrast with the system theoretic approach of thermodynamics by Haddad et al. [14]. In company with these recent developments, the present treatise is a systematic development and application of a new theory of motive force (power), long due after Carnot [15, 16]. The former faint ideas of the author, which go by the label "heuristic" [17] and "method of synthetic constraint" [18], are formally forged into a generalized formulation recognized as a natural tendency and hence perhaps may be regarded as a law of nature.

The crisis of totalitarian victory is, from the perspective of history, an awkward predicament characterized by intellectual sloth, lack of imagination, and wishful thinking [19]. It is well known that no science develops systematically from one single starting point according to a definite preconceived plan, but its development depends on practical considerations and proceeds more or less simultaneously along different lines, corresponding to the many ways of looking at the problems, and to the times and views of the investigator. Thus, science cannot attain its objective by direct means, but only gradually along numerous and devious paths, and therefore a wide scope is provided for the individuality of the worker [20].

Admittedly, if we do not succeed in solving a mathematical or physical problem, it is often because we have failed to recognize the more general standpoint from which the problem before us appears as a single link in a chain of related problems. This way to find generalized methods is certainly the most practical and the surest one, for he who seeks the method without having a definite problem in mind seeks in vain [21]. For reasons, in this present endeavor, it is preferable to choose some topical problems that are of common interest both to physicists and to the engineering community. This effort is inspired by the seminal call of Bejan through a letter [22] that appeared in the Journal of American Association of Physics Teachers (AAPT).

Nowadays, it has become a fashionable trend [23–25] to publish volumes of empirical material without any thesis or antithesis, such as figures, photographs, computer generated images, and essays on the observation that both natural and engineered systems exhibit a category of symmetry [26, 27]. In contrast, this monograph is a submission against such strategies that may eventually open up the vision of contemporary as well as the next generation of researchers. This, at any rate in my opinion, lack or even absence of figures accompanying the analysis or description, actually stimulates the abstract thinking process, which is eventually the key to the problem solving aspect.

The purview of this current script is to purport a commonality of a diverse view of observations. At the present state of human knowledge and affairs, such a unified exact description of everything on a general footing, both at the macroscopic as well as microscopic [28] levels, will only be poor, vague, and scanty. A theory that is too general is frequently too weak. There is a way out, which I frequently describe to my disciples as an approach from the "periphery to the center." We must not look at the intricacies of the objects and events at the outset; we will keep a habit of looking into simplicities out of complexities and thus concentrate on the outer aspects of the subject during introspection. We will go on adding details in succession until we are undone with a realistic solution to the problem at our disposal. Thus, we will be able to compare apples to oranges. Dwelling on this qualitative aspect, everything appears to be a ramification of a single principle and a unique perspective, which is the object of the present treatise.

This memoir is the faithful disposition of a discourse that I witnessed and withstood with pain and pleasure as a rational as well as an emotional being. Education is the manifestation of perfection already present in man. This study is a passage to that destiny of freedom: from bondage to spiritual faith, from spiritual faith to great courage, from courage to liberty, from liberty to abundance, from abundance to selfishness, from selfishness to complacency, from complacency to apathy, from apathy to dependency, from dependency to back to bondage again. Where and how [29] do I break the chain? The teleological perspective [30, 31] of the present work that tacitly follows in disguise is however not a theme of this treatise. The highest motto underlying the curtain of thorough scientific investigation is but a true aspiration for self-knowledge and self-realization [32], or at least the awakening and sharpening of human faculties already attributed to us. For any conceivable physical principle, there must be a corresponding counterpart of mental (psychological) principle, which in turn is a replica of a metaphysical (spiritual) principle. Our complete realization will actually mean an assimilation of a principle distinctly at these three different levels of human perception. One will then at least be able to rejoice in an added confidence in thought, speech, and action [33]. These exercises were part and parcel of the character of the founders of modern science [34, 35]. An earnest study habit [36] will enable the reader to attain a greater vision to see, which is attributed only at the elevated consciousness [37]. For example, the clairvoyant investigations [38] into the structure of matter carried out by theosophists Besant and Leadbeater was confirmed [39, 40] by the physicist Philips through experimentation and scientific reasoning. As a matter of passing mention, a reader can check the progress on his way to attainment: while in deep thought (meditation), in a single chance you are able to look up a topic from a book without consulting the index first. Accordingly, I have adopted the following principles in my research, with *noblesse oblige*.

Research requires curiosity, diligence, devotion, and aimful thinking [41, 42]. The Latin *mundo corde* describes it better. One can learn so much out of anything, if one can truly start with a blissful ignorance [43]. A perished thought is surely a germinating one. Many a sleepless thoughtful night can give birth to a resourceful dawn. Also, a researcher has first much to do with the overcoming of one's own inferiority complex [44, 45]. As such, one should publish a piece of work when it is even imperfect and incomplete than perfect and complete never. To start with, topics may be chosen with reference to some works of authorities on the subject field. Refinement, generalization, and/or dismissal of their findings could be found as a means of gaining confidence in the research progress.

There are at least two distinct ways in which a subject field can be developed. One is the "horizontal" expansion into the more remote fields intersected by the subject. Another is the "vertical" expansion, that is, a deepening of our present understanding (inception, conception, and the perception) that defines the province. A large number of contemporary workers continue to regard the field of classical thermodynamics as matured and saturated; that is precisely why such old and prevalent topics are picked up. There remains not only a merit in questioning the established point of view, but also the fact that a true research frontier is, quite often, a territory overlooked by the crowd [46]. For such reasons the classical and fundamental research is sought, so that we learn to answer the question "why" and not "how" alone [47]. In every inch of the work, a good balance between the case-specific subjective findings and the general objective reality [48] of a scientific query of general nature [49] is being established.

Regarding the research publication guideline, a piece of advice by Moran was followed [50]: (a) If the work is in the realm of theory, then what truly new insights or relations are achieved, and what is their importance? (b) If the endeavor is in the realm of engineering, then what is the contribution? (c) Does the development provide at least a picture book engineering pointing the way to a significant evolution in some aspect of engineering practice? Also, the present research constantly haunted a physical principle devoid of many computer produced tables and graphs [25]. It was the untiring motivation of the current investigation that the purpose of the computing is insight but not numbers [51].

In my stride I am blissfully aware of the very presence of my masters who preached me to inculcate discipline, method, scholarship, taste, and style up to an adorable personality, for my own survival and succor. For such reasons, it is perhaps not untimely and unprovoked to support the view [52]: "It is recorded that Sancho Panza, when he saw his famous master charge into the windmills, muttered in his beard something about relative motion and Newton's Third law. Sancho was right: the windmills hit the master just as hard as he hit them."

Some peers may rate my presentation as rushed, unwise, and inordinately pretentious and abstract. Perhaps it is. But it is too late now to escape the influence of my masters who taught me that prudence is a rich old maid courted by incapacity. Should they pay heed to the warnings of William Blake?

"When all their Crimes, their Punishments, their Accusation
of Sin,
All Jealous Revenges, Murders, hiding of Cruelty
in Deceit
Appear only at the outward Sphere of Visionary Space and
Time
In the shadows of Possibility, by Mutual Forgiveness for
Evermore
And in the Visions & in the Prophecy, that we may Foresee
& Avoid
The terrors of Creation, Redemption & Judgment."

Durgapur, India
Guru Purnima: July 2013

Achintya Kumar Pramanick

References

1. Pramanick, A.K.: Philosophy of nature. Reflection magazine, pp. 2–5. R.E. College, Durgapur (1990–1991)
2. Pramanick, A.K.: Natural philosophy of thermodynamic optimization. Doctoral thesis (Unpublished), Indian Institute of Technology, Kharagpur (2007)
3. Pramanick, A.K.: Equipartition of Joulean heat in thermoelectric generators. In: Rocha, L.A.O., Lorente, S., Bejan, A. (eds.) Constructal Law and the Unifying Principle of Design. Springer, New York (2013)
4. Klein, M.J.: Carnot's contribution to thermodynamics. Phys. Today **27**, 23–28 (1974)
5. Thomson, W.: Mathematical and Physical Papers-I. Cambridge University Press, London (1882)
6. Bejan, A.: Advanced Engineering Thermodynamics, p. 50. Wiley, New York (1997)
7. Feyerabend, P.: Against Method, p. 27. Verso, London (1978)
8. Truesdell, C.: The Tragicomedy of Classical Thermodynamics. CISM, Udine, Courses and Lectures, No. 70. Springer, New York (1983)
9. Truesdell, C.: Rational Thermodynamics, pp. 1–57. Springer, New York (1984)
10. Bejan, A.: Shape and Structure, from Engineering to Nature. Cambridge University Press, Cambridge (2000)
11. Bejan, A., Zane, J.P.: Design in Nature: How the Constructal Law Governs Evolution in Biology, Physics, Technology, and Social Organization. Anchor Books, New York (2013)
12. Bejan, A.: Advanced Engineering Thermodynamics, p. 807. Wiley, New York (1997)
13. Lieb, E.H., Yngvason, J.: The physics and mathematics of second law of thermodynamics. Phys. Rep. **310**, 1–96 (1999)
14. Haddad, W.M., Chellabonia, V.S., Nersesov, S.G.: Thermodynamics: A Dynamical Systems Approach. Princeton University Press, New Jersey (2005)
15. Mendoza, E. (ed.): Reflections on the Motive Power of Fire. Dover, New York (2005)
16. Kestin, J. (ed.): The Second Law of Thermodynamics-I. Dowden, Hutchinson and Ross, Pennsylvania (1976)
17. Pramanick, A.K., Das, P.K.: Heuristics as an alternative to variational calculus for optimization of a class of thermal insulation systems. Int. J. Heat Mass Transf. **48**, 1851–1857 (2005)
18. Pramanick, A.K., Das, P.K.: Method of synthetic constraint, Fermat's principle and the constructal law in the fundamental principle of conductive heat transport. Int. J. Heat Mass Transf. **50**, 1823–1832 (2007)

19. Bridgman, P.W.: A challenge to physicists. J. Appl. Phys. **13**, 209 (1942)
20. Planck, M.: A Survey of Physical Theory (trans: Jones, R., Williams, D.H.), p. 82. Dover, New York (1993)
21. Hilbert, D.: Mathematical Problems. Arch. Math. Phys. **3**(1), 44–63, 213–237 (1901)
22. Bejan, A.: Engineering advances on finite-time thermodynamics. Am. J. Phys. **62**, 11–12 (1994)
23. Bejan, A.: Advanced Engineering Thermodynamics, p. x. Wiley, New York (1997)
24. D' Le Alembert, J.R.: Nouvelles expériences sur la résistance des fluids. Lambert, Paris (1777) (in French)
25. Truesdell, C.: Six Lectures on Modern Natural Philosophy, pp. 100–101. Springer, New York (1966)
26. Weyl, H.: Symmetry. Princeton University Press, New Jersey (1983)
27. Feynman, R.: The Character of Physical Law, pp. 84–107. MIT Press, Cambridge (1985)
28. Čápek, V., Sheehan, D.P.: Challenges to the Second Law of Thermodynamics: Theory and Experiment. Springer, New York (2005)
29. Dubrovsky, D.: The Problem of the Ideal: The Nature of Mind and Its Relationship to the Brain and Social Medium (trans: Stankevich, V.). Progress, Moscow (1983)
30. Burtt, E.A.: The Metaphysical Foundation of Modern Science. Dover, New York (2003)
31. Esbenshade Jr., D.H.: Relating mystical concepts to those of physics: some concerns. Am. J. Phys. **50**, 224–228 (1982)
32. Gough, A.E.: The Vaisheshika Aphorism of Kanada: With Comments from Upasakara, of Sankara Misra and the Vivriti of Jaya Narayana Tarkapanchana. Motilal Banarsidass, New Delhi (1976)
33. Whorf, B.L.: Language, Thought, and Reality: Selected Writings. In: Carroll, J.B. (ed.). MIT Press, Cambridge (1956)
34. Leibniz, G.W.: Discourse on Metaphysics and the Monadology (trans: Montgomery, G.R.). In: Chandler, A.R. (ed.). Dover, New York (2005)
35. Newton, I.: Newton's Philosophy of Nature: Selections from His Writings. In: Thayer, H.S. (ed.). Dover, New York (2012)
36. Dvivedi, M.N.: The Yoga Sutras of Patanjali. Motilal Banarsidass, New Delhi (2000)
37. Besant, A.: A Study in Consciousness. Theosophical Publishing House, Adyar (1999)
38. Besant, A., Leadbeater, C.W.: Occult Chemistry. Theosophical Publishing House, Adyar (1951)
39. Phillips, S.M.: Anima: Remote Viewing of Subatomic Particles. Theosophical Publishing House, Adyar (1996)
40. Phillips, S.M.: ESP of Quarks and Superstrings. New Age International, New Delhi (2005)
41. Wiener, N.: Invention: The Care and Feeding of Ideas. MIT Press, Cambridge (1993)
42. Munk, M.M.: My early aerodynamic research—thoughts and memories. Ann. Rev. Fluid Mech. **13**, 1–7 (1981)
43. Hopfield, J.J., Feinstein, D.I., Palmar, R.G.: 'Unlearning' has a stabilizing effect in collective memories. Nature **304**, 158–159 (1983)
44. Besant, A.: Thought Power: Its Control and Culture. Quest Books, Illinois (1988)
45. James, W.: Talks to Teachers on Psychology and to Students on Some Life's Ideals. Dover, New York (2001)
46. Bejan, A.: Advanced Engineering Thermodynamics, p. xv. Wiley, New York (1997)
47. Lythcott, J.: "Aristotelian" was given the answer, but what was the question? Am. J. Phys. **53**, 428–432 (1985)
48. Born, M.: Natural Philosophy of Cause and Chance, p. 218. Dover, New York (1964)
49. Thompson, B.: An enquiry concerning the source of heat which is excited by friction. Philos. Trans. R. Soc. Lond. **88**, 80–102 (1798)
50. Moran, M.J.: On second-law analysis and the failed promise of finite-time thermodynamics. Energy **23**, 517–519 (1998)
51. Hamming, R.: Numerical Methods for Scientist and Engineers, p. 3. Dover, New York (1987)
52. Den Hartog, J.P.: Mechanics, p. v. Dover, New York (1961)

Acknowledgments

> *A good many times I have been present at gatherings of people who by the standards of the traditional culture are thought highly educated and who have with considerable gusto been expressing their incredulity at the illiteracy of scientists. Once or twice I have been provoked and have asked the company how many of them could describe the second law of thermodynamics. The response was cold; it was also negative. Yet I was asking which is about the scientific equivalent of: "Have you read a work of Shakespeare's?"*
>
> C. P. Snow

The investigation reported in this memoir spans for about two and half decades in three different continents, viz., India (National Institute of Technology Durgapur, Jadavpur University, and Indian Institute of Technology Kharagpur), the United States of America (Louisiana State University, Baton Rouge), and Europe (Technische Universität Chemnitz, Germany). I connote my gratitude to the authorities of these educational institutions and a few other academias where I professed, as they offered me a great opportunity for research study with their subsequent financial support in terms of fellowships, scholarships, and salaries. I would like to express a token of appreciation to the laboratory incharges and their supporting staff for allowing me to work round the clock without any hindrance or interference. The librarians and coordinating staff of these institutes provided exemplary assistance and cooperation by furnishing me with an exceedingly large volume of documents through interlibrary loan services.

I do not know how to acknowledge my mentors Prof. Sukamal Ghosh from National Institute of Technology (NIT) Durgapur, Prof. Achintya Kumar Mukhopadhyay and Prof. Swarnendu Sen from Jadavpur University (JU), Prof. Srinath V. Ekkad from Louisiana State University (LSU), Prof. Prasanta Kumar Das from Indian Institute of Technology (IIT) Kharagpur, and Prof. Karl Heinz Hoffmann from Technische Universität (TU) Chemnitz. Any word of thankfulness will perhaps be a misnomer. I am much obliged that they agreed to supervise my researches and allowed me to pursue simultaneously a second vocation in new, fundamental, and challenging areas of contemporary interests of mine all along in thermodynamics. It is their meticulous surveillance and close tutelage that enabled me to conclude this philosophical work in my lifetime. I recall how my supervisors persistently motivated even during the difficult moments of my research. I cherish their kind support being my friend, philosopher, and guide in all ways.

IIT Kharagpur imparted a major impact on my research career. To begin with, I was a visiting fellow at Center for Theoretical Studies (CTS) for a short period. During the lean period between the submission of my doctoral thesis in January 2007, and until the acceptance of this proposed monograph on November 2012, my bosom friend Dr. Partha Pratim Bandyopadhyay from IIT Kharagpur continued to invite me for a collaborative work at CTS and thus kept my spirit kindled. The rest of the void in my creativity was partially mitigated by Prof. Gautam Biswas, Director of Central Mechanical Engineering Research Institute (CMERI) Durgapur, India.

Professor Karl Heinz Hoffmann from Institute of Physics, Computational Physics group, TU Chemnitz was a magnanimous host in my research career. I enjoyed inculcating the freedom of ideas in his company spread over almost one and a half years. Working with him was more of a pleasure than a privilege alone. His critical but constructive criticism persuades me to adopt an active research career in physics and particularly in quantum thermodynamics. Miss Angelique Gaida, secretary to Prof. Hoffmann, was very kind and prompt in replenishing the materials required for research, including a very large number of books from time to time.

With the advent of a rapid communication system, a number of international academic personalities provided me with their authored treatises and/or the reprint of articles completely free of cost and finally lent their attention for consultation in a number of occasions. Among a host, in alphabetical order pertinent to my work, I thankfully acknowledge the generosity of Prof. Alexis De Vos, Universiteit Gent, Belgium; Dr. Anatoly Tsirlin, Program Systems of Russian Academy of Sciences, Russia; Prof. André Thess, Technische Universität Ilmenau, Germany; Prof. Bernard Howard Lavenda, Università Camerino, Italy; Prof. Bjarne Andresen, University of Copenhagen, Denmark; Prof. Dick Bedeaux, Norwegian University of Science and Technology, Norway; Prof. Elias Panayiotis Gyftopoulos, Massachusetts Institute of Technology, USA; Prof. Gian Paolo Beretta, Universitá de Brescia, Itally; Prof. Hans Ulrich Fuchs, Zurich University of Applied Sciences at Winterthur, Switzerland; Prof. Ingo Müller, Technische Universität Berlin, Germany; Prof. Jeffery Lewins, University of Cambridge, UK; Prof. Jeffrey M. Gordon, Ben-Gurion University of the Negev, Israel; Prof. Jincan Chen, Xiamen University, China; Prof. Massoud Kaviany, University of Michigan, Ann Arbor, USA; Prof. Michel Feidt, Institut National Polytechnique de Lorraine et Universite Henri Poincare, France; Prof. Peter Salamon, San Diego State University, USA; Prof. Richard Stephen Berry, University of Chicago, USA; Prof. Signe Kjelstrup, Norwegian University of Science and Technology, Norway.

Old wine tastes better. A large number of seasoned friends, former students, and teachers continue to be on my stride. Residing well outside the give-and-take relationship, they flare up my tempo. I still bask in the support and encouragement rendered by Prof. Sumanta Acharya from LSU, Baton Rouge, USA. I could not escape the everlasting influence of Prof. Samir Kumar Saha, from Jadavpur

University, who introduced and finally enticed me into research on the fundamental frontiers of thermodynamics while I was a student of my first master's degree.

I forgot to mention that researchers are also emotional creatures. I am bonded to a large number of hail-fellow-well-mate peers through their cordiality and congeniality for sharing the moments of success and failure with equal ease and comfort and thus aptly creating a social environment for sustenance. My interest in spirituality, palmistry, and above all, openly being praised and criticized as well as reciprocating drew them nearer to my heart. Right from attending movie shows to gossiping in a tea stall or near a vending machine are simply filled with immortal and undefiable memories. Without this impulsive and unobtrusive environment I doubt I would have inducted the breath and the food for thought of my research. Accordingly, I engrave my commitment of faithfulness to a large of number of pals. With a responsible fear of missing to mention any one, I abstain from showcasing my ungratefulness.

On a sentimental note, I reveal and reiterate my tender feelings to my parents—my staunch well-wishers, brothers, sisters, and all other family members for their relentless spontaneous support and sacrifice of personal gratifications towards the realization of my own academic ambitions. The lion's share of my thanksgiving is due to my better half for her patient understanding and endless endurance. It was her pleasant duty to awake me whenever my alarm clock were fade up at times. She crosschecked several drafts of the main body of this work and assisted with the placement of references, figures, and the overall organization. However, for any inadvertent error, I bear the sole responsibility.

While conjuring up the past, I am amused at the awful moment when my spiritual Guru Swami Paramanada Maharaj charged my consciousness as if I got a major bash for the first time in life. I learned to some extent to live in this Vanity Fair otherwise being incorrigibly namby pamby and vulnerable in my approach. His uncalled for compassion (*kripa*) composed me to be totally ransacked and devastated out of my mental constitution (*samaskara*). His preaching of polydimensional geometry and to realize up to seven dimensions in this life itself is a continuous persuasion to submit this monograph. He is writing through my pen with this imperfect instrument like me. By the virtue of studentship alone, I took it for granted the enormous encouragement and buttress provided by Prof. Adrian Bejan, Duke University, USA. I started communicating with him about his works through postal letters beginning in October 1997 and by August 2000 I was supposed to be his doctoral student. Somehow, this ambition could not get matured but I continue to be his pupil. This treatise is an inescapable magnetism of him.

An impetus and harbinger to this writing is a series of events beginning with a review of a book in August 2011 and finishing with a publication of a book chapter of mine in January 2013. In the process I became a fan of Springer. On the anvil of time after several communications, in November 2012, I consented to publish this monograph to my publishing editor Dr. Leontina De Cecco and coordinator Dr. Holger Schäpe. I cannot miss this opportunity to express my profound gratitude to Dr. Leontina and Dr. Schäpe, who so spontaneously, patiently, and cheerfully goaded, prodded, pushed, wheedled, and cajoled me into finishing in reasonable time, and above all, adopted my idea.

Although crafting the section on acknowledgments has the advantage of being ocular, it frequently suffers from the stigma that what is seen is only sensed. I sincerely apologize for any omission that might have slipped in oblivion.

Why art thou silent? Is thy love a plant
Of such weak fibre that the treacherous air
Of absence withers what was once so fair?
Is there no debt to pay, no boon to grant?

W. Wordsworth

Contents

Nomenclature

Letters

a	Acceleration of a particle
a	Arbitrary complex number
a	Constant [Eq. (2.43)]
a_m	Constant
A	Area
A	Cross-sectional area of a fluid column
A	Cross-sectional area of a leg of a thermoelectric module
A	Dimensionless parameter [Eq. (3.35)]
A	Heat exchanging surface area
A	Total heat exchanger surface area
ΔA	Elemental surface area
A_H	Hot end heat exchanger surface area
A_l	Area of the lower strip
A_L	Cold end heat exchanger surface area
A_u	Area of the lower strip
b	Constant [Eq. (2.43)]
b	Dimensionless parameter [Eq. (3.33)]
b	Nonzero complex number
B	Dimensionless parameter [Eq. (3.35)]
Bi	Biot number [Eq. (3.6)]
Br_x	Local Brun number [Eq. (2.45)]
c	Constant heat capacity
c_i	Causal factors, $i = 1, 2, \ldots, n$ [Eqs. (1.5–1.7)]
c_i	Parametric constants, $i = 1, 2, 3, \ldots, 22$
C	Centroid
C	Constant [Eq. (1.48)]
C	Constant [Eq. (2.43)]
C	Convection term [Eq. (1.17)]
C	Dimensionless parameter [Eq. (4.35)]
C	Finite constant [Eqs. (5.41a, 5.41b)]
C_a	Constant [Eq. (1.1)]

C_A	Finite constant [Eq. (5.41d)]
$\bar{C}_a$	Constant
C_B	Finite constant [Eq. (5.41f)]
C_h	Constant [Eq. (1.14)]
C_i	Internal thermal conductance of the power plant
C_K	Finite constant [Eq. (5.41c)]
C_r	Constant [Eq. (1.4)]
$\bar{C}_r$	Constant
C_{ra}	Constant [Eq. (2.2b)]
C_{rv}	Constant [Eq. (2.1b)]
C_s	Constant [Eq. (1.3)]
$\bar{C}_s$	Constant
C_{sa}	Constant [Eq. (2.2a)]
C_{sv}	Constant [Eq. (2.1a)]
C_T	Constant [Eq. (2.31)]
C_Z	Finite constant [Eq. (5.41e)]
d	Diffusion like flow quantity
D	Diffusion term [Eq. (1.17)]
Ec	Eckert number
f	Function [Eq. (1.32)]
f	Function [Eq. (1.50)]
f	Function [Eq. (1.82)]
f	Function obtained from Blasius solution [Eq. (4.11)]
$\hat{f}$	Force field
f_b	Backward motivation [Eq. (1.2)]
f_f	Forward motivation [Eq. (1.2)]
F	Function of insulation thickness [Eq. (2.15a)]
F	Shorthand for an integrand [Eq. (3.23)]
F	Thrust on elemental fluid area [Eq. (4.53)]
F_a	Applied force [Eq. (1.8)]
F_b	Backward motivation [Eq. (1.1)]
F_e	Effective force [Eq. (1.12)]
F_f	Forward motivation [Eq. (1.1)]
F_H	Fraction of Joulean heat affecting high temperature heat source
F_i	Force of inertia [Eq. (1.10)]
F_k	Kinetic energy [Eq. (1.14)]
F_L	Fraction of Joulean heat affecting low temperature heat sink
F_p	Potential energy [Eq. (1.14)]
F_x	Pressure force in horizontal direction
g	Function [Eq. (1.82)]
g	Gravitational acceleration
g	Known function [Eq. (1.42)]
G_b	Backward motivation [Eq. (1.2)]
G_f	Forward motivation [Eq. (1.2)]

h	Approximate height of the fluid stream before and after hydraulic jump
h	Constraint [Eq. (1.44)]
h	Elemental length of fluid column
h	Local convective heat transfer coefficient
h	Wave length [Eq. (4.49)]
$\bar{h}$	Location of hydrostatic force
h_i	Heat transfer coefficient between fluid stream and cylindrical wall [Eq. (2.32a)]
h_i	Height of a fluid column, $i = 1, 2, 3, \ldots, n$
$\bar{h}_i$	Depth of center of pressure, $i = 1, 2, 3, \ldots, n$
h_L	Heat transfer coefficient at the extreme downstream [Eq. (3.3)]
h_P	Location of hydrostatic force [Eq. (4.35)]
$\bar{h}_P$	Location of hydrostatic pressure with reference to a pole
h_0	Heat transfer coefficient between insulation and ambient [Eq. (2.32a)]
h_1	Height of the fluid stream before hydraulic jump
h_1	Reference height
$h_{1,\min}$	Minimum height of the fluid stream before hydraulic jump [Eq. (4.47)]
h_2	Height of the fluid stream after hydraulic jump
$h_{2,\min}$	Minimum height of the fluid stream after hydraulic jump [Eq. (4.48)]
H	Depth of a gate
H	Height of isothermal fluid column
H	Orthogonal dimension
i	Arbitrary branching level
i	Number of rooms
i	Particular segment
$\hat{i}$	Unit vector
I	Area moment of inertia of a fluid stream
I	Electric current in a branched network
I	Flow current [Eq. (1.52)]
I	Functional [Eq. (1.32)]
I	Integral [Eqs. (1.35, 1.49)]
$\dot{I}$	Functional [Eq. (1.78)]
I_C	Moment of inertia with reference to centroid
I_i	Constant electric current in a branched network, $i = 0, 1, 2$
$\hat{j}$	Unit vector
J	Dimensionless group [Eq. (4.17)]
$\bar{J}$	Dimensionless physical parameter [Eq. (4.16)]
$\bar{J}_{\max}$	Upper ceiling of insulation volume [Eq. (3.32)]
J'_q	Local heat flux
J_x	Electrical current density vector along x-direction
k	Integral constraint [Eq. (1.42)]
k	Local conductivity of insulating material
$\hat{k}$	Unit vector
k_f	Thermal conductivity of fluid

k_i	Thermal conductivity, $i = 1, 2$
k_i	Thermal conductivity of insulating material [Eq. (2.32a)]
k_i'	Modified thermal conductivity, $i = 1, 2$ [Eq. (2.37c)]
k_w	Conductivity of cylindrical wall [Eq. (2.32a)]
k_w	Thermal conductivity of insulating material
K	Constant [Eq. (1.51)]
K	Thermal conductance
K_H	Thermal conductance of high temperature side heat exchanger
K_L	Thermal conductance of low temperature side heat exchanger
L	Lagrangian [Eq. (1.63)]
L	Length of a cylinder
L	Length of a flat plate
L	Length of the leg of a thermoelectric device
L	Thermodynamic distance
L	Wall length
$L_{\max}$	Maximum permissible length of a thermoelectric module [Eq. (5.10a)]
$L_{\min}$	Minimum permissible length of a thermoelectric module [Eq. (5.10b)]
$L_i L_i'$	Length of the ith thermoelectric module
m	Mass of a particle
m	Number of competing mechanisms [Eq. (3.44)]
m	Number of segments
M	Fixed point location
M_i	Parallel faced ith homogeneous medium, $i = 1, 2, 3, \ldots, N$
n	Exponent in heat transfer coefficient relation [Eq. (3.3)]
n	Index of power law [Eqs. (4.31, 6.3, 6.4)]
n	Nondimensional parameter [Eq. (2.14a)]
n	Number of horizontal parts of a fluid column
n	Number of passages at a certain branching level
n	Type of semiconductor material
$\dot{n}$	Population moving per unit time
$\dot{n}''$	Population per unit area and time
n_1	Real part of complex power law index
n_2	Imaginary part of complex power law index
N	Branching levels
N	Net efflux per unit volume [Eq. (1.27)]
N	Number of contiguous parallel faced homogeneous media
N	Number of rooms
Nu_x	Local Nusselt number [Eq. (2.43)]
p	Pressure on elemental fluid element [Eq. (4.54)]
p	Thermodynamic pressure
p	Type of semiconductor material
p_{atm}	Atmospheric pressure
p_i	Pressure of a fluid column, $i = 1, 2, 3, \ldots, n$
p_1	Reference pressure

P	Fixed location on an area
P	Moving point in two-dimensional plane
P	Pole
P	Power output of the engine [Eq. (6.13)]
ΔP	Maximum pressure difference
$\bar{P}$	Dimensionless power [Eq. (6.20)]
P_i	Arbitrary point in a flow field
P_j	Arbitrary point in a flow field
P_r	Real part of dimensionless power [Eq. (6.35)]
Pr	Prandtl number
$\hat{q}$	Heat flux field
q'	Heat transfer rate per unit length [Eqs. (3.5, 3.21)]
q'_{constant}	Heat transfer rate per unit length of constant thickness profile [Eq. (3.39)]
$q'_{\min}$	Minimum heat transfer rate per unit length [Eq. (3.29)]
$q'_{*\min}$	Minimum heat transfer rate per unit length [Eq. (3.10)]
q'_{taper}	Heat transfer rate per unit length of tapered profile [Eq. (3.34)]
q''	Local heat flux [Eq. (3.4)]
Δq	Local heat transfer rate
Δq_i	Heat transfer from ith segment, $i = 1, 2, 3, \ldots, m$
Q	Heat
Q	Volume of a fluid stream
ΔQ	Form of local heat transfer rate [Eq. (2.3)]
$\dot{Q}$	Heat transfer rate duty
$\dot{Q}'$	Unsteady heat transfer rate
$\dot{Q}_H$	Heat transfer rate from the high temperature source
$\dot{Q}_H$	Steady state heat transfer rate supplied by the heat source [Eq. (6.8)]
$\dot{Q}_H^*$	Heat flow rate to the hot end [Eq. (5.12)]
$\dot{Q}_{*H}$	Heat flow rate to the hot end [Eq. (5.23)]
$\dot{Q}_{HC}$	Steady heat transfer rate between work producing compartment and high temperature side [Eq. (6.5)]
$\dot{Q}'_{HC}$	Unsteady heat transfer rate between work producing compartment and high temperature side [Eq. (6.3)]
$\dot{Q}_i$	Steady bypass heat leak through the machine structures [Eq. (6.7)]
$\bar{\dot{Q}}_i$	Dimensionless bypass heat leak
$\dot{Q}_J$	Joulean heat transport rate [Eq. (5.14)]
$\dot{Q}_k$	Conducted heat transport rate [Eq. (5.13)]
$\dot{Q}_L$	Heat transfer rate from the low temperature sink
$\dot{Q}_L$	Steady state heat rejected at the heat sink [Eq. (6.9)]
$\dot{Q}_{LC}$	Steady heat transfer rate between work producing compartment and low temperature side [Eq. (6.6)]
$\dot{Q}'_{LC}$	Unsteady heat transfer rate between work producing compartment and high temperature side [Eq. (6.4)]

$\dot{Q}_{LCE}$	Steady state heat released to the heat sink by endoreversible heat engine
r	Outer radius of a cylindrical wall
r	Volumetric ratio of the high to the low conductive material
R	Electrical resistance
R	Flow resistance [Eq. (1.52)]
R	Rate term [Eq. (1.15)]
R	Resistance to heat flow
R	Universal gas constant
$\hat{R}$	Vectorial quantity
$\Delta\hat{R}$	Small distance
$\bar{R}$	Total conductive and convective resistance [Eq. (3.7)]
R_i	Resistance at ith branching level
R_i	Time independent parallel resistors, $i = 1, 2$
R_l	Electrical resistance of the lower strip
Re_L	Reynolds number at the extreme downstream
R_s	Electrical resistance of any strip element
R_u	Electrical resistance of the upper strip
Re_x	Local Reynolds number
s	Arc length of the path of light
s	Slenderness ratio
s_{opt}	Optimum slenderness ratio
S	Entropy
S	Scalar quantity
S	Shape factor
S	Source term [Eq. (1.15)]
ΔS	Change in scalar quantity [Eq. (1.21)]
S_{cup}	Entropy of the cup
$\dot{S}_{\mathrm{gen}}$	Entropy generation rate
$\bar{\dot{S}}_{\mathrm{gen}}$	Uniform entropy generation rate [Eq. (3.43)]
S_i	Scalar quantity
S_j	Scalar quantity
S_{room}	Entropy of the room
S_s	Shape factor of a rectangular strip
S_{universe}	Entropy of the universe
S_φ	Source term [Eq. (1.31)]
t	Local insulation thickness
t	Passage time of light [Eq. (1.45)]
t	Time
$\bar{t}$	Length-based averaged wall thickness [Eq. (2.16)]
$\bar{t}$	Length-based averaged wall thickness [Eq. (3.2a)]
t_{1l}	Optimal insulation thickness distribution [Eq. (2.21)]
t_{2l}	Optimal insulation thickness distribution [Eq. (2.22)]
t_{3l}	Optimal insulation thickness distribution [Eq. (2.23)]

t_{opt}	Optimal insulation thickness [Eq. (2.35)]
t_{opt}	Optimal insulation thickness [Eq. (3.28)]
t_{taper}	Tapered insulation profile [Eq. (3.33)]
t_w	Thickness of the wall [Eq. (2.32a)]
t_*	Optimal insulation thickness distribution [Eq. (3.9)]
T	Average absolute temperature of a thermoelectric module
T	Temperature distribution function [Eq. (5.1)]
T	Thermodynamic temperature
T	Wall temperature variation
T_0	Ambient temperature
T_0	Temperature at the bottom surface of the plate
T_0	Wall temperature at $x = 0$
T_1	Wall temperature variation [Eq. (2.40)]
T_∞	Free stream temperature
ΔT	Applied temperature gap across a thermoelectric module
ΔT	Constant thermal potential difference
ΔT	Local temperature gradient
ΔT	Maximum temperature difference
ΔT_r	Relative temperature drop term [Eq. (2.44)]
T_f	Local fluid stream temperature
T_h	Real temperature component of the high temperature side [Eq. (6.1)]
T_H	Heat source temperature
T_H	Highest temperature of rooms
T_H	High temperature side temperature
T_{HC}	High temperature level at which the device actually receives the heat
T_{HC}	Transient temperature of the working fluid at the hot end [Eq. (6.1)]
T_{HO}	Time-averaged temperature of the working fluid at the hot end
T_i	Temperature, $i = 1, 2$
T_i	Temperature of the ith room, $i = 1, 2, 3, \ldots, N$
T_l	Real temperature component of the low temperature side [Eq. (6.2)]
T_L	Heat sink temperature
T_L	Low temperature side temperature
T_L	Lowest temperature of rooms
T_L	Wall temperature at $x = L$
T_{LC}	Low temperature level at which the device actually rejects the heat
T_{LC}	Transient temperature of the working fluid at the cold side [Eq. (6.2)]
T_{LO}	Time-averaged temperature of the working fluid at the cold end
T_w	Interfacial wall temperature [Eqs. (3.14a, 3.14b)]
u	Velocity component along the flat plate
u_i	Velocity of light in ith medium, $i = 1, 2, 3, \ldots, N$
U	Overall heat transfer coefficient
U	Total internal energy of the system
UA	Overall thermal conductance [Eq. (6.16)]
U_H	Overall heat transfer coefficient of high temperature side heat exchanger

U_L	Overall heat transfer coefficient of low temperature side heat exchanger
U_∞	Free stream velocity
v	Control volume size
v	Fixed upper volume of the strip
v	Specific volume of the fluid
v	Velocity component normal to the flat plate
v	Velocity of a particle
$\hat{v}$	Velocity vector
Δv	Form of elemental insulation volume [Eq. (2.3)]
v_l	Volume of the lower strip
v_u	Volume of the upper strip
v_1	Reference specific volume
V	Approximate velocity of fluid stream before and after hydraulic jump
V	Potential difference [Eq. (1.52)]
V	Voltage in a branched network
V	Volume of a fluid element
V	Volume of an insulating material
V_1	Velocity before hydraulic jump
V_2	Velocity after hydraulic jump
$\bar{V}$	Nondimensional insulation volume [Eq. (2.24b)]
$\hat{V}$	Vector flux
ΔV	Elemental insulation volume
ΔV	Voltage drop
V_i	Voltage in a branched network, $i = 1, 2$
ΔV_l	Potential drop of the lower strip
ΔV_u	Potential drop of the upper strip
V_x	Component of a vector flux
V_y	Component of a vector flux
V_z	Component of a vector flux
w	Width of the control volume
w	Width of the finite fluid column
W	Wall width
$\dot{W}$	Work output rate
x	Heat exchanger allocation ratio
x	Longitudinal coordinate direction
x	Orthogonal direction
x	Side of a fluid element of parallelogram shape
x	Variable
Δx	Linear dimension
x_i	Abscissa, $i = 0, 1, 2$
x_i	Constant, $i = 0, 1, 2$
x_i	Roots of a quartic equation, $i = 1, 2, 3, 4$ [Eq. (6.41)]
x_{opt}	Optimal heat exchanger allocation ratio
X	Longitudinal dimension of a rectangular block

X	Column vector of extensive variables
X_q	Conjugate driving force
y	Orthogonal direction
y	Side of a fluid element of parallelogram shape
y	Transformed heat exchanger allocation ratio [Eq. (6.46)]
y	Variable
y	Vertical coordinate direction
Δy	Linear dimension
y_i	Arbitrary function in x_i for $i = 0, 1, 2$
y_i	Constant, $i = 0, 1, 2$
y_i	Ordinate, $i = 0, 1, 2$
Y	Comparison function in x [Eq. (1.33)]
Y	Lateral dimension of a rectangular block
z	Figure of merit of a thermoelectric module [Eq. (5.39a)]
z	Orthogonal direction
z	Parameter
z	Transformed heat exchanger allocation ratio [Eq. (6.50)]
$\bar{z}$	Transformed heat exchanger allocation ratio [Eq. (6.58)]
Δz	Linear dimension
z_i	Roots of a cubic equation, $i = 1, 2, 3$ [Eq. (6.50)]

Greek Symbols

α	Dummy variable [Eq. (3.18)]
α	Inclination of a fluid element with the horizontal
α	Seebeck coefficient of the material
β	Dummy variable [Eq. (3.18)]
$\bar{\chi}$	Total magnitude of all dissipative forces [Eq. (3.44)]
χ_i	Arbitrary design variable, $i = 1, 2, 3, \ldots, m$ [Eq. (3.44)]
δ	Dimensionless parameter [Eq. (2.26b)]
δ	Thickness of heat exchanging medium
δ_T	Thermal boundary layer thickness
Δ	Dimensionless parameter [Eq. (2.27c)]
Δ	Infinitesimal difference
ε	Correction factor [Eq. (2.41)]
ε	Effectiveness of heat exchanging equipment
ε	Parametric constant [Eq. (1.33)]
ε_H	Effectiveness of high temperature side heat exchanger
ε_L	Effectiveness of low temperature side heat exchanger
ϕ	Degree of irreversibility [Eq. (6.10)]
ϕ	Intensive property
ϕ_i	Angle of the light ray with the normal at ith medium, $i = 1, 2, 3, \ldots, N$ [Eqs. (1.47, 1.48)]
Φ	Aggregate integral [Eq. (3.23)]

Φ	Extensive property
γ	Cost of unit conductance
γ	Specific weight of a fluid column
γ_H	Cost of unit conductance of high temperature side heat exchanger
γ_L	Cost of unit conductance of low temperature side heat exchanger
Γ_ϕ	Coefficient of diffusion
η	Arbitrary function in x
η	Similarity variable [Eq. (3.12)]
η	Thermal efficiency of the engine [Eq. (6.15)]
η_r	Real part of the engine efficiency [Eq. (6.72)]
κ	Thermal conductivity of the material
κ_e	Electrical conductivity
κ_l	Lattice thermal conductivity
κ_0	Constant thermal conductivity of a thermoelectric element
λ	Dimensionless parameter [Eqs. (5.6b, 5.15)]
λ	Lagrange multiplier
λ	Numerical and dimensional factor [Eqs. (2.1a, 2.1b)]
$\hat{\lambda}$	Accommodating factor [Eq. (1.1a)]
λ_i	Constant, $i = 1, 2$ [Eqs. (1.57, 1.58)]
λ_i	Numerical and dimensional factor for ith segment, $i = 1, 2, 3, \ldots, m$ [Eqs. (2.1a, 2.1b)]
Λ	Dimensionless parameter [Eq. (5.6a)]
Λ_1	Parametric group [Eq. (2.19)]
Λ_2	Parametric group [Eq. (2.27a)]
Λ_3	Parametric group [Eq. (2.29a)]
Λ_4	Parametric group [Eq. (3.30a)]
μ	Numerical and dimensional factor [Eqs. (2.2a, 2.2b)]
μ_i	Numerical and dimensional factor for ith segment, $i = 1, 2, 3, \ldots, m$ [Eqs. (2.2a, 2.2b)]
v	Kinematic viscosity
θ	Angle of applied force
θ	Dimensionless temperature [Eq. (5.4a)]
θ	Included angle between two nonparallel sides of a fluid element
θ	Nondimensionalized fluid temperature [Eq. (3.12)]
θ_1	Angle of incidence [Eqs. (2.37b, 2.37c)]
θ_2	Angle of refraction [Eqs. (2.37b, 2.37c)]
θ_*	Dimensionless temperature distribution without Thomson heat [Eq. (5.19)]
θ^*	Dimensionless temperature distribution with Thomson heat consideration [Eq. (5.8)]
ρ	Density of fluid element
ρ	Electrical resistivity of the material
ρ_e	Effective resistivity of any strip element
ρ_i	Density of a fluid column, $i = 1, 2, 3, \ldots, n$
ρ_l	Electrical resistivity of the lower strip element

ρ_u Electrical resistivity of the upper strip element
ρ_0 Constant electrical resistivity of a thermoelectric element
σ Electrical conductivity or reciprocal of electrical resistivity of the material
$\dot{\sigma}$ Local entropy production rate
$\dot{\sigma}_{\mathrm{EoEP}}$ Local entropy production rate with equipartition of entropy production
$\dot{\sigma}_{\mathrm{EoF}}$ Local entropy production rate with equipartition of forces
$\dot{\sigma}_{\mathrm{opt}}$ Optimum local entropy production rate
τ Ratio of high temperature to the low temperature
τ Temperature ratio spanned by heat source and sink [Eq. (6.18)]
τ Thomson coefficient of the material
τ_h Intermediate temperature ratio [Eq. (6.18)]
τ_o Steady state temperature ratio spanned by working fluid [Eq. (6.18)]
τ_{opt} Optimized temperature range of the working fluid
ω Oscillating periodic frequency of thermal wave [Eqs. (6.1, 6.2)]
ξ Dimensionless leg length of the thermoelectric device [Eq. (5.4b)]
ξ Dimensionless length of the flat plate [Eq. (3.2b)]
ξ_* Location of maximum temperature without Thomson heat [Eqs. (5.21, 5.22)]
ξ^* Location of maximum temperature with Thomson heat [Eqs. (5.9, 5.11a, 5.11b)]
ψ Numerical and dimensional factor [Eq. (2.3)]
ζ Dimensionless heat transfer rate [Eq. (5.20)]

Subscripts

C Centroid
constant Uniform wall thickness distribution
cup Pertaining to the cup
e Quantities of electrical origin
EoEP Equipartition of entropy production
EoF Equipartition of forces
gen Rate of entropy generation
H Pertaining to the high temperature side
HC Transient quantities at the hot end
HO Time-averaged quantities at the hot end
i Index
i Number of competing dissipating mechanisms
i Number of segments considered in a continuous fluid stream
J Quantities related to component of Joulean heat
l Lower strip
l Quantities of lattice thermal origin
L Quantities related to low temperature sink
LO Time-averaged quantities at the cold side

m	Dummy variable [Eq. (3.44)]
max	Maximum
min	Minimum
min	Minimum with conjugate formulation [Eq. (3.29)]
$*$min	Minimum with nonconjugate formulation [Eq. (3.10)]
n	Integer number of partition considered in a finite length of fluid element
n	Parameter determining wall temperature curvature
n	n-type material
opt	Optimum
p	p-type material
P	Reference pole
q	Heat
r	Real part of a complex quantity
room	Pertaining to the room
s	Arbitrary strip
t	Time-averaged quantities
$\bar{t}$	Averaged thickness based quantities
taper	Tapered wall thickness distribution [Eq. (3.33)]
u	Upper strip
universe	Pertaining to the universe
x	Along x-coordinate direction
x	Flow direction
1	Refers to wall
2	Pertains to insulation
$*$	Optimum wall thickness distribution [Eq. (3.9)]
$*$	Quantities pertaining to without Thomson heat consideration
δ_T	Quantities based on thermal boundary layer thickness

Superscripts

atm	Atmospheric pressure
–	Averaged dimensionless quantity
–	Location of hydrostatic force
–	Transformed quantities
$*$	Quantities pertaining to Thomson heat consideration
0	Constancy of physical parameter

Symbols

$\langle\rangle$	Averaged quantity
Δ	Change in value

Abbreviations

EoEP	Equipartition of entropy production
EoF	Equipartition of forces
EoTD	Equipartition of temperature difference
EGM	Entropy generation minimization
ETD	Equal thermodynamic distance
FTT	Finite-time thermodynamics
PM	Power maximum

Chapter 1
Introduction

We must gather and group appearances, until the scientific imagination discerns their hidden laws, and unity arises from variety; and then from unity we must reduce variety, and force the discovered law to utter its revelations of the future.

W. R. Hamilton

There is one basic cause for all effects.

G. Bruno

Every cause produces more than one effect.

H. Spencer

All human knowledge thus begins with intuitions, proceeds then to concepts, and ends up with ideas.

I. Kant

A mathematical theory is not to be considered complete until you have made it so clear that you can explain it to the first man you meet on the street.

D. Hilbert

For when propositions are denied, there is an end of them, but if they bee allowed, it requireth a new worke.

F. Bacon

In this chapter, it is appropriate to begin with a brief overview of mathematical and physical principles to provide a coherent and self-contained account of the works that follow in subsequent chapters. The chief objective is to propose a concise physical theory of thermodynamics pertaining to the nature of motive force. The current emphasis is on now identifying the mechanisms and system components that are responsible for the optimum shape, structure, and performance of the system. Geometric form is another name for the macroscopic organization of the various parts (solid, fluid) of the heterogeneous flow system. What is being optimized in order to improve the global performance (power, efficiency, irreversibility, resistance, cost, etc.) is not nearly as important as how the imperfections of the flow system (resistances, entropy generation, etc.) must be distributed and balanced against each other for the evolution of the system. The method of analysis adopted here has a predominant basis on physical understanding of the underlying principle, which goes by the ideal of natural philosophy.

A. K. Pramanick, *The Nature of Motive Force*, Heat and Mass Transfer,
DOI: 10.1007/978-3-642-54471-2_1, © Springer-Verlag Berlin Heidelberg 2014

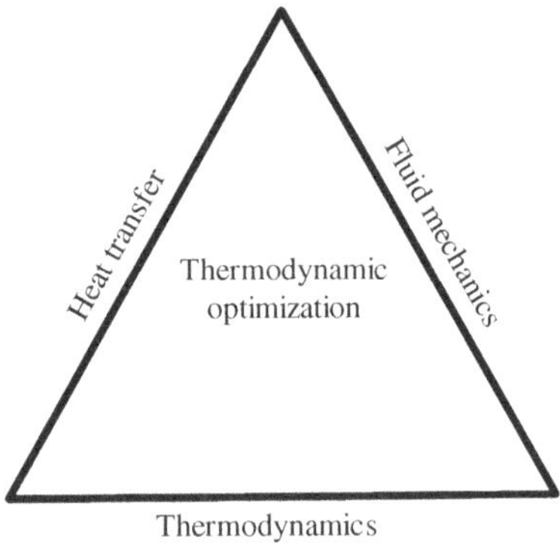

Fig. 1.1 Evolution of thermodynamic optimization

1.1 Motivation

We form our notions through the interpretation of interrelated observations. This method may rightly be called "Natural Philosophy," a word still used for physics at the Scottish universities [1]. As long as natural philosophy exists, its ultimate highest aim will always be the correlation of various physical observations into a unified system, and where possible, into a single formula. This is an arena inspired primarily by curiosity rather than necessity. Challenges from the cross-disciplinary areas with the confluence of heat transfer [2–21], fluid mechanics [22–37], and thermodynamics [38–64] are worthy to be called as thermodynamic optimization [65–67], which is the basis of the present investigation under the limelight of natural philosophy. This is schematically presented in Fig. 1.1.

During my undergraduate studies in the Mechanical Engineering discipline, at National Institute of Technology Durgapur, India, one piece of composition by the Count of Rumford (Benjamin Thompson) had left a hard-to-wipe-out-impression in my mind to the process of understanding (inception, conception, and the perception) in the vast field of learning [68]:

"It frequently happens, that in the ordinary affairs and occupations of life, opportunities present themselves of contemplating some of the most curious operations of nature; and very interesting philosophical experiments might often be made, almost without trouble or expense, by means of machinery contrived for the mere mechanical purposes of the arts and manufacturers. I have frequently had occasion to make this observation; and I am persuaded, that a habit of keeping the eyes open to every thing that is going on in the ordinary course of the business of the life has oftener led, as it were by accident, or in the playful excursions of the imagination, put into action by contemplating the most common appearances, to useful doubts and sensible schemes for investigating and improvement, than all the more intense meditations of philosophers, in the hours expressly set apart for study."

Constantly haunted by this inspiration, the present investigation is devoted to the study of the fundamental aspects of thermofluid science. While reporting the results of this present investigation, I was heavily charged with the influence of Truesdell [69]:

"Our colleagues in the professionalized natural sciences often look upon as quaint antiquarians if not reactionaries from the radical right. Just as the university has changed from a center of learning to a social experience for the masses, so research, which began as a vocation and became a profession, has sunk to a trade if not a racket. We cannot fight the social university and mass-produced research. Both are useful—useful by definition, since they are paid, if badly. But we must not allow social university to destroy learning, and the trade of research to take away our right and capacity to think. Society demands and pays for commercial art and canned music, but the employees in these industries do not hold themselves up as ideals towards which every painter and composer should strive. In contrast, the organized trade of science, not yet sufficiently distinct to boast its indifference to the old- fashioned individual ways, decries them as antiquated and evil, and seeks to strangle the vocation of science, dredging the public pocket as well as filling the public press with the triumphs of massive teams of 'experts' lulled by the costly blink of binary numbers by the billion. Soon, perhaps, small children will skip from door to door, begging dimes for digits. No one will deny that the giant brains in obedience to teams of little ones can do things undreamt by our fathers. What has not been shown is any change at all in the requirements for the kind of science the scientists of the past created. Granted that trade science can pour out in a day more tables of calculation and curves of experimental data; that Newton could have inspected in a life time, I see no evidence that the kind of science Newton did, the science that has given, ultimately, the swarms of scientific ants the ground under their anthills, can be done in any other way than he did it. Today we all ride in trade-produce autos, but society does not jeer at athletes who run a race with their old-fashioned feet, just as runners run in ancient Athens. The athletes and the artists are allowed to pursue their vocations, indeed, supported in them, in the midst of a beehive society. Likewise, a quiet corner must be found for the learned and for the creators, even in the modern university. While not surrendering to the trade science, the natural philosopher must not form misguided and suicidal snobbery attack to it. Trade science is invincible, but it need not remain an enemy, for between trade science and science as a vocation, there is more misunderstanding than real conflict."

Once a noble and revered science, for example, think of Leonardo da Vinci, Sadi Carnot, and the airplane builders during World War II, engineering is now taken for granted. Everywhere we look, from university campus politics to the Nobel Prize, engineering ranks either too low or not at all at the ladder of respect. The engineering reality is a lot brighter [70]. The improvers of the mechanical arts were neglected by biographers and historians, from a mistaken prejudice against practice, as being inferior in dignity to contemplation; and even in the case of men such as Archytas (an ancient Greek philosopher) and Archimedes, who combined practical skill with scientific knowledge, the records of their labors that have reached our time give but vague and imperfect accounts of their mechanical inventions, which are treated as matters of trifling importance in comparison with their philosophical speculations. The same prejudice, prevailing with increased strength during the middle ages, and aided by the prevalence of the belief in sorcery, rendered the records of the progress of practical machines, until the end of the fifteenth century, almost a blank. These remarks apply, with peculiar force, to the history of those machines called "prime movers" [71]. That is why Rankine, "the engineer and co-founder of classical thermodynamic," is almost never mentioned by philosophers. There were several books written by the so-called great mathematicians with a pedantic proclamation that Fermat's last theorem is impossible to prove. But the German engineer Max Munk did the job by the grace of his intellect, which puzzled the mathematical society for a century [72, 73]. Prandtl, a mechanical engineer and co-founder of fluid mechanics, did not know much mathematics; this was the unbiased observation of one of his great students, von Karman. But Prandtl proposed the boundary layer theory that serves the mathematical province of partial differential equation [74]. The inertia [75–77] of matter is so much in vogue from time immemorial that we neither question its origin nor its true nature, rather we feel more comfortable in taking it as obvious all along. A tribute to Amitava Ghosh, ex-director of Indian Institute of Technology Kharagpur, once again a mechanical engineer by vocation, who did unravel it recently with simplicity and excellence [78] by employing Mach's principle [79]. On the same boat and with the same spirit, this monograph will pursue the engineering quest in the methodology of natural philosophy.

1.2 Aim and Scope

This body of work chiefly addresses some interrelated fundamental problems of contemporary interest, especially thermodynamicists. The laid down approach paves a way to the grassroots of education—that is, how our thought should be modified and to what extent we should be dependent on calculating machines. Nevertheless, the results obtained will find its industrial application too apart from the far-reaching consequences of the conceptualizing processes of fundamental importance from the bird's viewpoint as is practiced in natural philosophy. For some reasons, classically unsolved problems from the thermofluid science discipline have been addressed and exact solutions provided by analytical means, and with the aid of physical reasoning and our proposed law of motive force. This monograph however is not a collection of problems, rather the ample applications of a newly recognized law of nature. The choice of problems is diversified but coherent.

The central theme of the investigation lies with the establishment of qualitative similarities with everything that we come across. We begin with the development of the theory of thermal insulation systems. This class of thermal systems is vast, complex, and numerous. To start with, we address the practical purpose of a thermal insulation system. One common and generally accepted view is that thermal insulation as a system prevents two bodies with surfaces of different temperatures from coming into direct thermal contact. Figure 1.2 schematically illustrates three different types of classical thermal systems, which qualify as thermal insulations in accordance with the general definition.

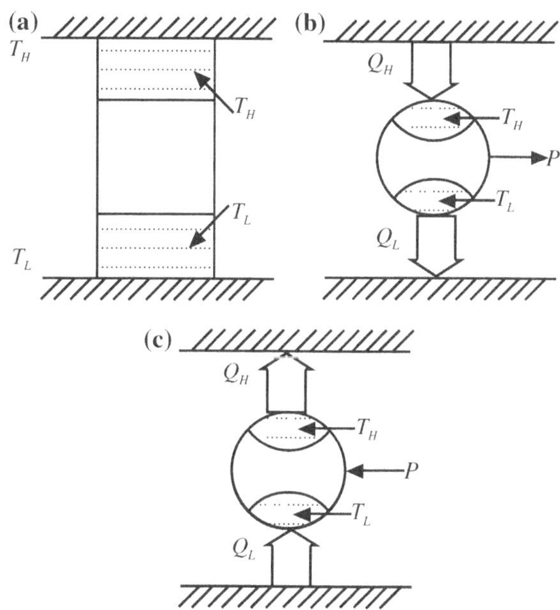

Fig. 1.2 Three classical examples of thermal insulation systems preventing two surfaces of different temperatures from coming into direct thermal contact: **a** conducting layer; **b** heat engine; **c** refrigerator

The typical case of a low thermal conductivity material sandwiched between two bodies with temperatures T_H (high temperature) and T_L (low temperature) is shown in Fig. 1.2a. Although the two bodies may communicate mechanically, the hot body is in direct contact with a body of equal temperature T_H. The same is the case for the cold body with temperature T_L. A power plant and refrigeration plant operating in cycles, while in thermal communication with two heat reservoirs, are shown in Fig. 1.2b, c, respectively. In these examples, the hot reservoir points toward the hot end of the cycle, while the cold reservoir is in contact with the cold end of the same cycle. So although two bodies with temperatures T_H and T_L communicate to exchange heat, they do not make direct contact due to the presence of cyclic devices. Therefore, heat engines and refrigerators function as thermal insulation systems in the broad sense of the definition of thermal insulation. Three important conclusions emerge from this discussion. The first is the discovery of qualitative resemblances among diverse events. The second is that thermal insulation, as a class, is varied, numerous, and complex. The third feature is that heat transfer from bodies with temperatures T_H and T_L are not necessarily constant [80].

The chapters are arranged as the problems have evolved toward the direction of greater complexity. Each chapter is introduced with appropriate references to acquaint the reader with the profoundness of the problems. This chapter is devoted to outline the newly recognized law of nature: the law of motive force. The chapter also provides the physics and mathematics background required to consult the monograph.

In Chap. 2, we study the conductive heat transport system and solve classically unsolved problems using our proposed new law of motive force. Based on this law, a general optimization methodology is inculcated to replace a body of variational formulations of some problems. Without taking recourse to rigorous variational formulation it is demonstrated from the physical perspective of the problem that for such a class of optimization problems a truly optimum exists. While seeking a basis for analogies among physical theories, the relation among many physical theories and laws is established. The chapter concludes with a solution of nonconjugate formulation of some conductive–convective heat transfer problems.

Chapter 3 is an outcome of the physical solution methodology introduced earlier to the solution of conductive and nonconjugate heat transfer problems. Our proposed law of nature is further exploited in some conductive–convective conjugate heat transfer problems so as to solve and generalize the classical problem of Pohlhausen's with Hartee's velocity profile completely analytically with reference to the engineering application of thermal insulation system design. This optimization methodology based on physical principle is computationally more advantageous and even more amenable than the formal variational formulation of the same problem. Finally, a method of intersecting asymptotes is employed for a meaningful exercise of any optimization methodology.

To start with Chap. 4, we closely examine the constructal theory and Fermat's principle considering point-to-point and volume-to-volume flow with reference to a fluid flow system in the limelight of our proposed law. We include the effect of gravity in constructal formulation and predict the hydraulic jump phenomenon theoretically for the first time. We also speculate the fundamental geometric building block in a shearing fluid flow. Finally, we observe a category of equipartition principle among naturally organized phenomenological systems vis-à-vis optimized systems.

In Chap. 5, we consider the thermodynamic analysis of a thermoelectric generator, which is regarded as a natural heat engine. It has forbearance with the theory of insulation design as the heat engine can be thought of a category of the thermal insulation system. From the viewpoint of constructal theory, we directly look into the shape and structure of the thermoelectric generator. We considered the finite-time irreversibility of the external heat transport mechanism. Internal irreversibility is attributed to the simultaneous heat and current flow. An investigation of Joulean heat distribution resulted in a fractal-like deterministic structure of a cascaded thermoelectric generator assembly. The same concern of Joulean heat distribution led to the optimal heat exchanger allocation and deduced the most appropriate constraint for the heat exchanger inventory allocation problem. In this example too, we confirm the presence of our proposed law.

The closing Chap. 6 is a sequel of natural heat engine. Here, we study the model of a more realistic heat engine. We consider the temperature of the working fluid at the hot end and the cold side to be the complex quantities. Newton's law of cooling for heat transport irreversibility is replaced by a generalized power law.

The relaxation effect in heat transfer is included. Both the bypass heat leak and internal irreversibility are taken into consideration. Finally, the engineering quest of optimal heat exchanger allocation is attended for maximum power output as the self-optimized system ensues. We examine invariably the proposed new law that is at work.

1.3 General Background

1.3.1 Law of Motive Force

In order to control rather than to manipulate nature, a thorough understanding of the phenomena occurring in nature, which we call natural phenomena, can never be denied. The systematic study of nature lies in the fact that, however diversified may be the character of nature and whatever may be the natural or artificial processes, there exist at least some similarities among them. It is therefore natural to seek a single law that governs nature. This yields nothing but some qualitative analysis of the natural phenomena, however, the quantitative formulations may vary from one field to another.

Nature itself is inactive and does not allow any change to take place, whatsoever, until and unless it is obliged and hence remains in a constrained condition till it resumes its "original state." Queries therefore arise about how the activity of nature out of this notion of hyper-inactivity can be accommodated. Before to set a satisfactory answer to this query we must be clear about the following phenomenon. When an "effort" tries to produce a change in any natural system, the system carries an innate tendency to be noncooperative with the "effort." We shall call the former effort as "forward motivation" and latter tendency as "backward motivation." In 1990, I recognized that an evolution of any system occurs out of a conflict between two opposing forces and these nomenclatures were proposed [81]. These forward and backward motivations are but the motive force [82] of a system. They are one and the same thing, only manifested differently on the basis of our objective. In thermodynamics, after the celebrated work of Sadi Carnot [83] in 1824, the philosophy of motive has not been investigated and inculcated adequately for its universal applications and to discover its far reaching consequences.

When the forward motivation is zero, we contemplate that the system is in its "apparent original state." Equality of forward motivation with the backward motivation implies that the system is in "dynamical equilibrium." A further differential increment of the forward motivation causes an infinitesimal deviation of the system from its original state. So now, it is well understood that for a particular change to take place, forward motivation will have to be provided in a particular way, which we describe as a "trap." In sum, the **law of motive force** can be phrased as:

Every motive force is self-contradictory in its existence.

This goes tacitly unnoticed with every system we study. It was not recognized and formulated as a fundamental law of nature before this work. It can be surveyed in every field of human knowledge.

There are many ways of viewing this newly recognized law of motive force. The first and foremost to my taste, choice, and preference is the mental visualization of any event that we recognize as thought experiment [84–89]. However, in the engineering realm we may wish to assimilate our concept through practical examples. The concept of motive force can be easily perceived by considering a simple reversible electrolytic cell [90]. By a reversible electrolytic cell we mean a cell where a reversal of the direction of the current flowing through it causes the chemical reactions taking place in it to proceed in the opposite direction. The "electromotive force" is none other than the "motive force" discussed here in this context. In our terminology, "electromotive force" itself is a "forward motivation" for the system, which is here the electrolytic cell for the "motive" to deliver the current. The "backward motivation" is offered by the "internal resistance" of the cell. Now, electromotive force can easily be related to the thermodynamic properties of the system, such as internal energy [90]. Further, it can be shown that the electromotive force of a reversible cell is a measure of the free energy change of the process taking place in the cell and this fact can be readily utilized to determine the activities and activity coefficients [91]. However, in the application of this new law of motive, we may not necessarily seek the relationship of motive force to those of thermodynamic properties.

The formation of pattern [92–126] in natural and in artificial systems demands the presence of some kind of motive force that may be viewed as a deviation from local thermodynamic equilibrium [127] resulting from differences in measurable thermodynamic properties or nonproperties of a system. It is only important to recognize the unit by which we can measure these motive forces. For example, we may consider the freezing of a certain liquid. In order to freeze that liquid, one has to extract the latent heat of freezing. On the other hand, this amount of extracted energy would be sufficient to heat up by some degrees of temperature T in some scale. Now, if we rapidly cool the fluid, it does not freeze at the stipulated freezing point, rather we can supercool the fluid quite substantially. In particular, if we supercool the fluid by the critical amount of T degrees before it starts freezing, the latent heat released on freezing is sufficient to reheat the resulting supercooled substance above the melting point. Hence, freezing would occur without the necessity of further heat extraction at a very fast rate. We can now express a motive force in such a fashion that it can be valid for any arbitrary system. The same formalism can be applied to very different situations without making any modification in the fundamental nature of competition among different entities. Only, we have to recognize the forward and backward motivation on the basis of the definition of our system and purpose. In what follows, though we have coined the tern "motive force," it will not be employed in a conventional manner [82]. From this discussion it turns out that any shape, structure, and process emerges out a competition among space, time, and matter [128–131].

Based on the above qualitative essence, in this section, we provide the glimpses of quantitative formulations of this newly admitted form of the law of motive force. Details are worked out in subsequent chapters with particular examples. From the thought experiment, it is to be realized that the same motive of a system is manifested as two different opposing tendencies of a system, such as forward motivation and backward motivation. In principle, it turns out to be a fact that it represents a category of "conservation principle" and competition of general nature [132–134]. Thus, we mathematically assert that

$$F_f + \hat{\lambda} F_b = C_a \tag{1.1}$$

where F_f and F_b amount to be the forward motivation and backward motivation, respectively, and C_a is a system and purpose-specific constant. The way the forward motivation and backward motivation have been defined, in general, may represent the competition among dissimilar quantities. The "accommodating factor" $\hat{\lambda}$ is thus both the numerical and dimensional factors for the consistency and dimensional homogeneity of Eq. (1.1). From the physical understanding of the relation (1.1), we recognize that F_f and F_b are finite positive quantities. Thus, we may write as

$$F_f = f_f^2 = G_f \text{ and } F_b = f_b^2 = G_b. \tag{1.2}$$

Further, from the physical basis of Eq. (1.1), we realize that increase in forward motivation leads to the decrease in backward motivation. Thus, Eq. (1.1) can also be alternatively framed as

$$G_f - \hat{\lambda} G_b = C_s. \tag{1.3}$$

In situations when one of the competing components of motive force runs to a constancy, Eq. (1.1) can be recasted as

$$\frac{F_f}{\hat{\lambda} F_b} = C_r. \tag{1.4}$$

Equations (1.1), (1.3), and (1.4) are capable of accommodating the thermodynamic theory of fluctuation [135] on a causal basis [136]. In that case, constants C_a, C_s, and C_r fluctuate around the mean values of $\bar{C}_a$, $\bar{C}_s$, and $\bar{C}_r$, respectively. Equations (1.1), (1.3), and (1.4) take a revised form in order as

$$F_f + \hat{\lambda} F_b = C_a(c_1, c_2, \ldots, c_n), \tag{1.5}$$

$$G_f - \hat{\lambda} G_b = C_s(c_1, c_2, \ldots, c_n), \tag{1.6}$$

and

$$\frac{F_f}{\hat{\lambda} F_b} = C_r(c_1, c_2, \ldots, c_n) \tag{1.7}$$

where c_i are the "causal factors" for $i = 1, 2, \ldots, n$. In summary, it can be pointed out that the law of motive force as represented by Eq. (1.1) spells out the first law of thermodynamics when both the forward and backward motivation pertains to energy [137]. By definition of forward and backward motivation, they each belong to the revelation of the second law of thermodynamics [138].

The purview and purpose of the present treatise is to testify the law of motive force for diversified classes of problems in the field of heat transfer, fluid mechanics, and thermodynamics. But the law of motive force so enunciated here belongs to a fundamental law of nature and hence other fields of exact and nonexact sciences should find their ready application and verification. Thus, it motivates us to outline some of the fundamental laws and principles of classical mechanics [139–166] in the framework of this newly acknowledged law of motive force. At the outset, we identify that forward motivation, backward motivation, as well as the motive force are the urges of a system depending on its objectives. Hence, these quantities resemble the force-like [167–175] entities. In our discussion, we use the term "mass" as a quantitative measure of "inertia" and by "inertia" we qualitatively mean the tendency of a system of doing what it was doing. We may measure mass by swinging an object in a circle at a certain speed and measuring how much force we need to keep it in a circle. In this way we find a certain quantity of mass for every object. Now, the momentum of a particle is a product of its mass and velocity. Thus, Newton's second law of motion can be expressed as [176]

$$F_a = \frac{\mathrm{d}}{\mathrm{d}t}(mv) = m\frac{\mathrm{d}v}{\mathrm{d}t} = ma \tag{1.8}$$

where F_a is the applied force causing a change in velocity v on mass m of a particle over a time span t on a linear path. We rewrite Eq. (1.8) as

$$F_a - ma = 0. \tag{1.9}$$

We can define another force called "force of inertia" F_i by the equation

$$F_i = -ma. \tag{1.10}$$

With this arrangement, Newton's second law of motion is reformulated as

$$F_a + F_i = 0. \tag{1.11}$$

Seemingly nothing is gained, since the intermediate relation (1.10) introduces merely a new terminology to the negative magnitude of product of mass times acceleration. It is precisely this apparent triviality that makes d'Alembert's principle [177] such an ingenious invention and at the same time so open to misunderstanding and distortion. The importance of the relation (1.11) lies in the fact that it is more than a reformulation of Newton's second law of motion. It is the expression of a principle: the law of motive force. It is known that the vanishing of a force in Newtonian mechanics implies a state of equilibrium. By this device a dynamical problem is reduced to a phenomenon of statics. Now, we define "effective force" F_e as

$$F_e = F_a + F_i = F_a - ma. \tag{1.12}$$

Thus, for an infinitesimal displacement d principle of virtual work [178] can be expressed as

$$F_e d\cos\theta = (F_a - ma)d\cos\theta = 0 \tag{1.13}$$

where θ is the angle of application of the force. It is well recognized that the central theme of classical mechanics is the principle of virtual work. This criterion immediately leads to the more special principle of stationary potential energy [179]. By means of Legendre transformation [180] the principle of virtual work yields the principle of complementary energy [181], which is a generalization of Castigliano's theorem [182]. Hamilton's principle [183] is derived from the principle of virtual work with the introduction of the concept of inertial force. Hamilton's principle in turn yields Lagrange's equations of motion [184] directly. Further, Newtonian equations of motion can be obtained from a Galilean relativistic point of view [185]. In conclusion, we observe that the law of motive force formulated by Eq. (1.1) is the same as d'Alembert's principle represented by Eq. (1.11), where clearly F_a is the forward motivation and F_i is the backward motivation. Both in Lagrange's equation as well as in Hamilton's principle it is explicit that they belong to the conservation of principle of energy, which is a corollary to the law of motive force, where the forward motivation F_k is the kinetic energy and backward motivation F_p is the potential energy, and thus we obtain

$$F_k + F_p = C_h \tag{1.14}$$

where C_h is a constant of a conservation principle demanded by the formulated law of motive force expressed in Eq. (1.1).

1.3.2 Conservation Principle

The presence of any kind of motive force and its counterparts, such as forward motivation and backward motivation, is best realized for a possible flow situation of any kind. We describe such a control volume engaged in flow as a flow field [186, 187] to hint that any physical quantity can gain a motive force within its proximity and at the same time it may also lose its original motive. The incoming physical quantity is labeled as influx and the outgoing physical entity as efflux to mimic the flux and force-like [167–175] quantities that are at least somewhat similar to those in electricity and magnetism [188, 189]. In general, such a flow process is time-dependent and represents a fierce competition among space, time, and matter. For such a flow to exist there must be a source S of some kind. The source pours into the control volume and experiences a rate of accumulation or depletion R of some physical quantity. The source can be treated as a cause and the rate of accumulation or depletion as an effect. For a perfect balance (equilibrium) between R and S we can write

$$R = S \tag{1.15}$$

which can be rewritten as

$$R - S = 0. \tag{1.16}$$

Equation (1.16) behaves in the same manner as Eq. (1.3). Now, the term R has a history: it is the outcome of the competition between a slower process and a faster process. The slower process is termed as diffusion D and the faster process is called as convection C. In nature, generally diffusion-like slower processes emerge first and then faster processes like convection. In this way, diffusion can be treated as cause and convection as effect. A perfect balance (equilibrium) between this cause C and effect D towards making up the term R will result in

$$C = D. \tag{1.17}$$

Equation (1.17) is arranged as

$$C - D = 0 \tag{1.18}$$

to fit into the form (1.3). Now, we realize that in general the flow process evolves out of severe competition among C, D, R, and S. Eliminating the common right-side quantity between Eqs. (1.16) and (1.18) one can obtain

$$R + C = D + S. \tag{1.19}$$

In general, Eq. (1.19) can be treated as a qualitative form of conservation principle where paring of the quantities is not important as long as equality between both the sides is valid. In Eq. (1.19), it is realized that each quantity on any side has its corresponding counterpart to the other side of the equality.

For a quantitative solution to the problem, we are interested in implementing Eq. (1.19) through mathematical terms. Apart from abstract generalization, we consider now a fluid flow situation as a specific example with respect to a control volume and a local thermodynamic equilibrium concept [127]. In thermodynamics, usually it is necessary to make a distinction between those properties of a substance whose measure depends on the amount of the substance present and those properties whose measure is independent of the amount of material present. The former is called extensive property and the latter is known as intensive property. Let Φ represent any arbitrary extensive property of fluid for which there exists a corresponding intensive property ϕ such that they can be related with reference to a miniscule control volume of size Δv by the distributive measure of the form $\Phi = \iiint \phi\rho \mathrm{d}v$ where ρ is the local density of the fluid medium. So, $\rho\phi$ designates the amount of corresponding extensive property contained in a unit volume. Now, we can express the rate of change with respect to time t of the relevant property per unit volume as

$$R = \frac{\partial}{\partial t}(\rho\phi). \tag{1.20}$$

Next, we find a mathematical replacement of the diffusion and convection terms. Source term will remain as a fictitious term which will render the conservation principle into a mathematical equality. In principle, any physical law should be independent of the coordinate system. We will argue later for a possible natural shape of a fluid element. Without any loss of generality, we consider here a control volume as a three-dimensional volume element of infinitesimal linear dimensions Δx, Δy, and Δz along x, y, and z directions in an orthogonal rectangular Cartesian coordinate system forming an infinitesimal rectangular parallelepiped of volume $\Delta x\Delta y\Delta z$. Without recognizing the natural form of a material object, the other mathematical way of looking at the problem is coordinate transformation [190]. We imagine a scalar field [187] where two neighboring points $P_i(x, y, z)$ and $P_j(x + \Delta x, y + \Delta y, z + \Delta z)$ are situated by a separation of a small distance $\Delta\hat{R}$. Here, $\hat{R}$ represents a vectorial quantity such that $\Delta\hat{R} = \hat{i}\Delta x + \hat{j}\Delta y + \hat{k}\Delta z$ where $\hat{i}, \hat{j}$, and $\hat{k}$ are unit vectors along x, y, and z orthogonal coordinate directions, respectively. Any scalar quantity S has a measure of $S_i(x, y, z)$ at the point P_i and $S_j(x + \Delta x, y + \Delta y, z + \Delta z)$ at the point $P_j(x + \Delta x, y + \Delta y, z + \Delta z)$ such that $\Delta S = S_j(x + \Delta x, y + \Delta y, z + \Delta z) - S_i(x, y, z)$. Performing a Taylor series expansion of S_j around the point P_i and retaining only the first-order terms we obtain

$$\Delta S = \frac{\partial S}{\partial x}\Delta x + \frac{\partial S}{\partial y}\Delta y + \frac{\partial S}{\partial z}\Delta z. \tag{1.21}$$

Physically we meant a linear variation by which we realized Eq. (1.21). Since the left side of Eq. (1.21) is a scalar by definition, the right side also has to be a scalar quantity. But each term of the right side of Eq. (1.21) is a product of two components, one of which is a scalar component of a vector such as Δx, Δy, or Δz. Thus, Eq. (1.21) can be viewed as a dot product like

$$\Delta S = \left(\hat{i}\frac{\partial S}{\partial x} + \hat{j}\frac{\partial S}{\partial y} + \hat{k}\frac{\partial S}{\partial z}\right) \cdot \left(\hat{i}\Delta x + \hat{j}\Delta y + \hat{k}\Delta z\right). \tag{1.22}$$

The first term in the bracket of the left side in Eq. (1.22) is known as gradient of S and written as [191]

$$\text{grad}\, S = \hat{i}\frac{\partial S}{\partial x} + \hat{j}\frac{\partial S}{\partial y} + \hat{k}\frac{\partial S}{\partial z}. \tag{1.23}$$

By Eq. (1.23) we physically realize that the gradient of a scalar field gives rise to a vector field. For example, in heat transfer, the negative of the gradient of temperature distribution T yields a vector field know as heat flux field $\hat{q}$ such that $\hat{q} = -\text{grad}\, T$. So the term $-\text{grad}\, T$ complies with a motive force that is causing the flow of heat. In fluid mechanics, the negative of the gradient of pressure p produces a vector field recognized as force field $\hat{f}$ such that $\hat{f} = -\text{grad}\, p$. The quantity $-\text{grad}\, p$ yields a motive force resulting in the flow of fluid [192]. Here, a negative sign implies that decrease in magnitude of one quantity will amount to increase in the other.

With reference to the above-mentioned volume element $\Delta x \Delta y \Delta z$ in a rectangular Cartesian coordinate system, we will discuss now the behavior of a vector field. We consider any vector flux $\hat{V}$ with its components V_x, V_y, and V_z along x, y, and z directions, respectively, such that $\hat{V} = \hat{i}V_x + \hat{j}V_y + \hat{k}V_z$. The amount of influx at a position x in x direction across the area $\Delta y \Delta z$ is $V_x \Delta y \Delta z$ and efflux at $x + \Delta x$ in x direction over the same area $\Delta y \Delta z$ is $V_{x+\Delta x} \Delta y \Delta z$. The net efflux across these two parallel surfaces can be obtained by considering only linear variations, which means a Taylor series expansion up to first-order terms as

$$V_{x+\Delta x} \Delta y \Delta z - V_x \Delta z = \left(V_x + \frac{\partial V_x}{\partial x} \Delta x \right) \Delta y \Delta z - V_x \Delta y \Delta z = \frac{\partial V_x}{\partial x} \Delta x \Delta y \Delta z. \quad (1.24)$$

Counting similar contributions from two other directions y and z, respectively, we obtain

$$V_{y+\Delta y} \Delta z \Delta x - V_y \Delta z \Delta x = \left(V_y + \frac{\partial V_y}{\partial y} \Delta y \right) \Delta z \Delta x - V_y \Delta z \Delta x = \frac{\partial V_y}{\partial y} \Delta x \Delta y \Delta z \quad (1.25)$$

and

$$V_{z+\Delta z} \Delta x \Delta y - V_z \Delta x \Delta y = \left(V_z + \frac{\partial V_z}{\partial z} \Delta z \right) \Delta x \Delta y - V_z \Delta x \Delta y = \frac{\partial V_z}{\partial z} \Delta x \Delta y \Delta z. \quad (1.26)$$

Combining Eqs. (1.24), (1.25), and (1.26) one can get the net efflux per unit volume N as

$$N = \frac{\partial V_x}{\partial x} + \frac{\partial V_y}{\partial y} + \frac{\partial V_z}{\partial z}. \quad (1.27)$$

Again Eq. (1.27) can be casted as

$$N = \left[\hat{i} \frac{\partial}{\partial x} + \hat{j} \frac{\partial}{\partial y} + \hat{k} \frac{\partial}{\partial z} \right] \cdot \left[\hat{i} V_x + \hat{j} V_y + \hat{k} V_z \right]. \quad (1.28)$$

In Eq. (1.28) it is observed that the right side quantity is a dot product of two quantities and hence a scalar. So the left side magnitude N is also a scalar quantity and is known as divergence [193] of $\hat{V}$ and denoted by div $\hat{V}$. If $\hat{V}$ represents a heat flux, div $\hat{V}$ is the amount at which heat is emanating per unit volume from a point source. When the vectorial quantity $\hat{V}$ express a velocity, div $\hat{V}$ gives the amount at which fluid is originating at a point per unit volume. So we observe that due to the divergence operation a physical variable is losing its motivation.

In our discourse, we view any physical phenomenon as a link of cause and effect. By observing on what effects a cause produces, we will recognize motive forces. In mathematical terms, causes are treated as independent variables while effects are dependent variables. Any conservation principle employs a certain physical quantity as its dependent variable (effect) and implies that there must be a balance among various physical factors (causes) that influence the dependent variable (effect). From our discussions, it is learned that primary cause of flow is a slow diffusion-like [194–196] process which is gradient-dependent. So the primary origin of a flow-like quantity d scales with grad ϕ such that $d \sim$ grad ϕ and $d = \Gamma_{\phi}$ grad ϕ where Γ_{ϕ} is a scale factor known as coefficient of diffusion. Thus, the flow due to diffusion can be accounted as

$$D = \operatorname{div}\left(\Gamma_{\phi} \operatorname{grad} \phi\right). \tag{1.29}$$

On the other hand, convection [197–199] is a faster mode of transport than diffusion. So the convection-like faster process can be represented as

$$C = \operatorname{div}(\rho \phi \hat{v}) \tag{1.30}$$

where $\hat{v}$ stands for the velocity of transport. Combining Eqs. (1.19), (1.20), (1.29), and (1.30) we obtain

$$\frac{\partial}{\partial t}(\rho \phi) + \operatorname{div}(\rho \phi \hat{v}) = \operatorname{div}\left(\Gamma_{\phi} \operatorname{grad} \phi\right) + S_{\phi} \tag{1.31}$$

where S_{ϕ} signifies that the source term S is influenced by the nature of the physical variable ϕ. Equation (1.31) reveals that it represents a general nature of a conservation principle [200–202].

1.3.3 Variational Formulation

Let there exist a twice differentiable function $y = y(x)$ satisfying the conditions $y(x_1) = y_1$ and $y(x_2) = y_2$ which renders the integral

$$I = \int_{x_1}^{x_2} f(x, y, y')\mathrm{d}x \tag{1.32}$$

an extremum. Then in analytical terms of the variational [203–232] formulation of the problem we seek the differential equation in $y(x)$ that satisfies the integral equation (1.32). The constants x_1, y_1, x_2, y_2 are assumed to be known and f is a given function of the arguments x, y, y', which are twice differentiable with respect to any or any combination of them. We denote the function that minimizes or maximizes Eq. (1.32) by $y(x)$ and proceed to form the one-parameter family of comparison functions $Y(x)$ defined as

$$Y(x) = y(x) + \varepsilon\eta(x) \tag{1.33}$$

where $\eta(x)$ is an arbitrary differentiable function for which

$$\eta(x_1) = \eta(x_2) = 0 \tag{1.34}$$

and ε are the parameters of the family. The condition (1.34) ensures that $Y(x_1) = y(x_1) = y_1$ and $Y(x_2) = y(x_2) = y_2$; that is, all the comparison functions pass through the endpoint values. Geometrically, we mean one-parameter families of curves $y = Y(x)$ connecting the points (x_1, y_1) and (x_2, y_2). The optimizing arc $y = y(x)$ is a member of each family for $\varepsilon = 0$. The vertical deviation of any curve $y = Y(x)$ from the actual minimizing or maximizing arc is given by $\varepsilon\eta(x)$. Replacing y and y' in Eq. (1.32), respectively, by $Y(x)$ and $Y'(x)$ we form the integral

$$I(\varepsilon) = \int_{x_1}^{x_2} f(x, Y, Y')\mathrm{d}x \tag{1.35}$$

where, for a given function $\eta(x)$, this integral is clearly a function of the parameter ε. The argument Y' of the integral is provided through Eq. (1.33) by

$$Y' = Y'(x) = y'(x) + \varepsilon\eta'(x). \tag{1.36}$$

In view of Eq. (1.33), setting ε equal to zero is equivalent to replacing Y and Y', respectively, by y and y'. Thus, the integral (1.35) is an extremum with respect to the single variable ε. Then the necessary condition for an optimum, the vanishing of the first derivative of I with respect to ε, must hold for $\varepsilon = 0$, i.e.,

$$I'(0) = 0. \tag{1.37}$$

Employing the rule for the derivative of an integral with respect to a parameter, we obtain

$$\frac{\mathrm{d}I}{\mathrm{d}\varepsilon} = I'(\varepsilon) = \int_{x_1}^{x_2} \left(\frac{\partial f}{\partial Y}\frac{\partial Y}{\partial \varepsilon} + \frac{\partial f}{\partial Y'}\frac{\partial Y'}{\partial \varepsilon}\right)\mathrm{d}x = \int_{x_1}^{x_2} \left(\frac{\partial f}{\partial Y}\eta + \frac{\partial f}{\partial Y'}\eta'\right)\mathrm{d}x \tag{1.38}$$

from Eq. (1.35) with the aid of Eqs. (1.33) and (1.36). Since setting ε equal to zero is equivalent to replacing (Y, Y') by (y, y'), we have according to Eqs. (1.37) and (1.38)

$$I'(0) = \int_{x_1}^{x_2} \left(\frac{\partial f}{\partial y}\eta + \frac{\partial f}{\partial y'}\eta' \right) \mathrm{d}x = 0. \tag{1.39}$$

Integrating by parts the second term of this integral we obtain

$$I'(0) = \int_{x_1}^{x_2} \left[\frac{\partial f}{\partial y} - \frac{\mathrm{d}}{\mathrm{d}x}\left(\frac{\partial f}{\partial y'} \right) \right] \eta \mathrm{d}x = 0. \tag{1.40}$$

Since Eq. (1.40) must hold for all η, by employing the fundamental lemma of calculus of variations [233], we conclude that

$$\frac{\partial f}{\partial y} - \frac{\mathrm{d}}{\mathrm{d}x}\left(\frac{\partial f}{\partial y'} \right) = 0. \tag{1.41}$$

Equation (1.41) is the so-called Euler–Lagrange differential equation. If in addition to extremizing I, the required function $y(x)$ satisfies an integral constraint of the type

$$k = \int_{x_1}^{x_2} g(x, y, y')\mathrm{d}x \tag{1.42}$$

where $g(x, y, y')$ is a known function, then $y(x, \lambda)$ satisfies the new Euler–Lagrange equation

$$\frac{\partial h}{\partial y} - \frac{\mathrm{d}}{\mathrm{d}x}\left(\frac{\partial h}{\partial y'} \right) = 0 \tag{1.43}$$

where

$$h = f + \lambda g. \tag{1.44}$$

The parametric constant λ is known as Lagrange multiplier [234] and its value is determined by substituting the $y(x, \lambda)$ solution into the integral constraint (1.42). Solution of this equation is the basis for seeking the optimum. In our present endeavor in this work, we provide an alternative physical solution methodology that will not involve the solution of Euler–Lagrange equation. We employ the law of motive force for a system to solve a class of problems that may or may not admit variational forms. In principle, the law of motive force renders natural optima for all classes of problems [235].

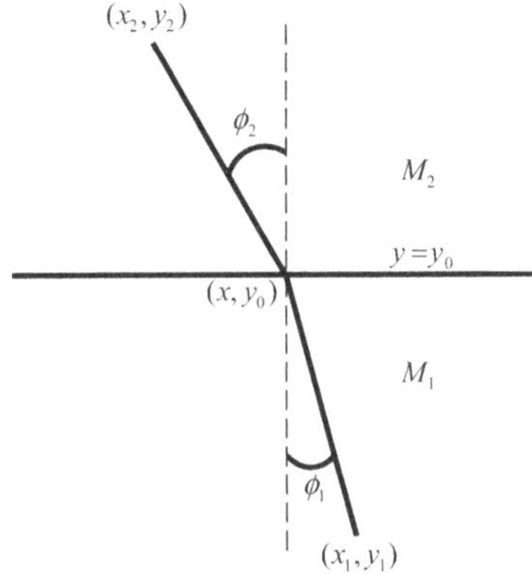

Fig. 1.3 Geometrical construction of Fermat's principle for two adjacent homogeneous optical media

1.3.4 Fermat's Principle

In geometrical optics, the principle of Fermat [236–238] states that the time elapsed in the passage of light between fixed points is an extremum with respect to possible paths connecting the points.

Now, in Fig. 1.3, we apply Fermat's principle to the passage of light from the point (x_1, y_1) in a homogeneous medium M_1 to the point (x_2, y_2) in another homogeneous medium M_2, which is separated from M_1 by the line $(x_1 < x_2)$. The respective velocities of light in the two media are u_1 and u_2. If we designate the point of intersection of an arbitrary two-segment path with $y = y_0$ as (x, y_0), the time for the passage of light would be

$$t = \frac{\sqrt{(x - x_1)^2 + (y_0 - y_1)^2}}{u_1} + \frac{\sqrt{(x_2 - x)^2 + (y_2 - y_0)^2}}{u_2}. \tag{1.45}$$

According to Fermat's principle, therefore, the actual path of the light is characterized by the value of x for which

$$\frac{\mathrm{d}t}{\mathrm{d}x} = \frac{x - x_1}{u_1\sqrt{(x - x_1)^2 + (y_0 - y_1)^2}} - \frac{x_2 - x}{u_2\sqrt{(x_2 - x)^2 + (y_2 - y_0)^2}} = 0 \tag{1.46}$$

or

$$\frac{\sin \phi_1}{u_1} = \frac{\sin \phi_2}{u_2} \tag{1.47}$$

where ϕ_1 is the angle between the normal to the interface $y = y_0$ and the path in M_1, ϕ_2 is the corresponding angle in M_2. The relation (1.47) is known as Snell's law [239] of refraction of light at the interface of two homogeneous media. Snell's

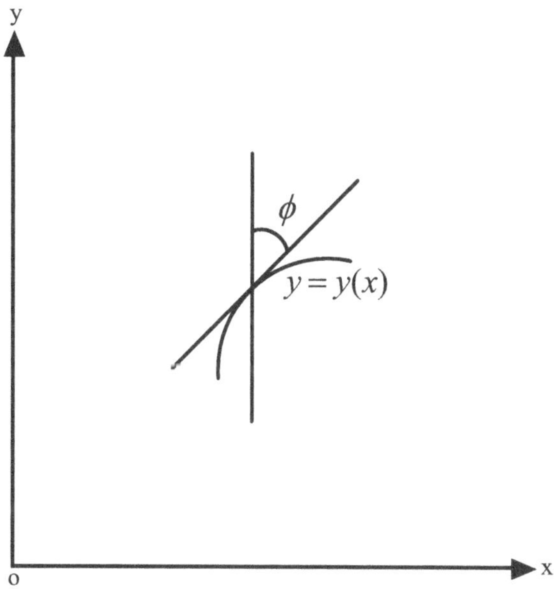

Fig. 1.4 Variational formulation of Fermat's principle for nonhomogeneous optical medium

law is experimentally well established. Equation (1.47) can be easily generalized for any set of N contiguous parallel-faced homogeneous media $M_1, M_2, \ldots, M_N$ as

$$\frac{\sin\phi_i}{u_i} = C \text{ for } i = 1, 2, \ldots, N \tag{1.48}$$

where ϕ_i is the angle that makes with the normal at the ith medium, u_i is the velocity in the ith medium, and C is a constant for any given light path.

In Fig. 1.4, we consider Fermat's principle applied to inhomogeneous medium. If the velocity of light is given by the continuous function $u = u(y)$, the actual light path connecting the points (x_1, y_1) and (x_2, y_2) is one that extremizes the time integral

$$I = \int_{(x_1,y_1)}^{(x_2,y_2)} \frac{\mathrm{d}s}{u} = \int_{x_1}^{x_2} \frac{\sqrt{1+y'^2}}{u}\mathrm{d}x \tag{1.49}$$

where ds is the elemental arc length of the path of the light. Equation (1.49) is valid even if $u = u(x, y)$. For $y(x)$ to be the optimum, the equation of the actual path of light must satisfy Euler–Lagrange equation (1.41) in which

$$f = \frac{\sqrt{1+y'^2}}{u(y)}. \tag{1.50}$$

After simplification we obtain

$$\frac{1}{u\sqrt{1+y'^2}} = K \tag{1.51}$$

where K is a constant. From Fig. 1.4 we read that $\sin\phi = \frac{1}{\sqrt{1+y'^2}}$. Thus, the equivalency of Eq. (1.51) with Eq. (1.48) is immediate. On account of the physical nature of Fermat's principle, it is to be noted that it is a point-to-point flow. Fermat's principle also fits into the framework of the law of motive force.

1.3.5 Constructal Law

Unlike point-to-point flow of Fermat's principle, the constructal law [240–253] deduces flow structures basically emanating out of point-to-area and point-to-volume flows. A generalized statement for the constructal law can be stated as [254, 255]: "For a finite-size open system to persist in time (to live), it must evolve in such way that it provides easier access to the imposed (global) currents that flow through it." Such a generalized law is referred as the fourth law of thermodynamics [254].

This theory resulted as a generalization of a class of access optimization problems. As an example, we consider the following flow problem [243]. Let us consider a finite-size geographical area A and a point M situated inside A or on its boundary, as shown in Fig. 1.5. Each member of the population living on A must travel between his point of residence $P(x, y)$ and the point M. The latter serves as a common destination for all the individuals who live on A. The density on this traveling population "that is, the rate at which individual must travel to M" is fixed and described by the parameter $\dot{n}''$ having the dimension of people (number) per unit area per unit time. This also means that the rate at which people are streaming into M is constrained, $\dot{n} = \dot{n}''A$. Now, we seek to determine the optimal cluster of paths that link the point P of area A with the common destination M, such that the time of travel required by the entire population is the minimum.

In a nutshell, the access optimization problem is how to connect a finite area (A) to a single point (M). It is to be noted that the area A contains an infinite number of points and every point must be taken into account when optimizing the access from A to M and vice versa. This problem is more complicated than the empirical game of connecting finite number of points ("many points") distributed over an area. The many-point problem can be solved on the computer using methods like random walk or Monte Carlo method, which are not theories [256].

There is an analogous problem of the fundamental access optimization issue known as Steiner's problem in mathematics. It states how to connect several points of a specified finite area by the shortest line [257–268]. An alternate to Steiner's problem is the constructal theory for volume-to-point access. With reference to

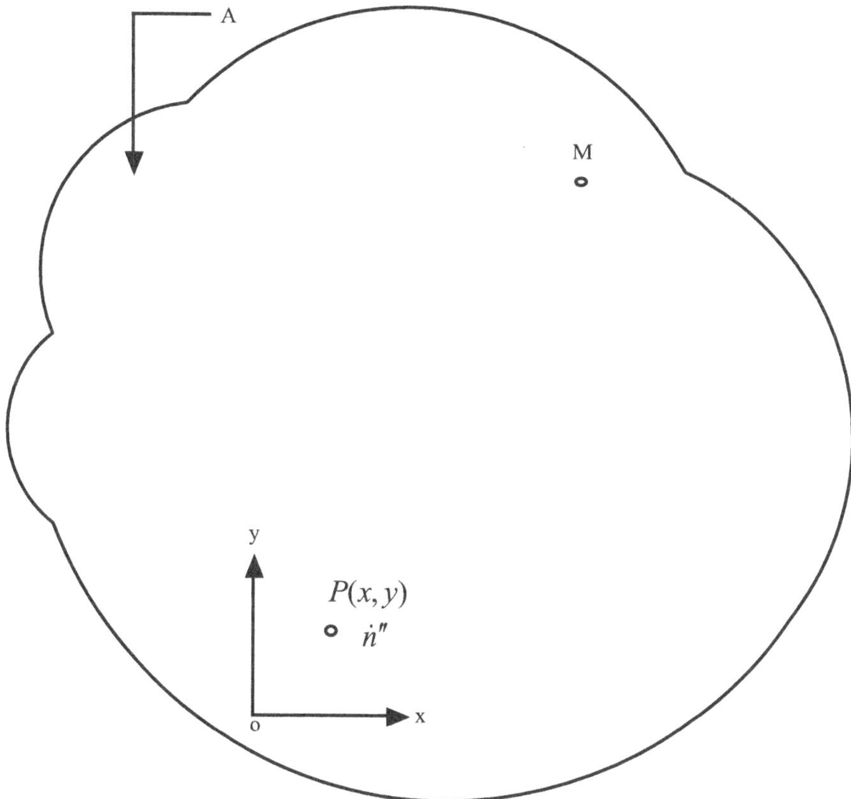

Fig. 1.5 Illustration of constructal theory as a point-to-area access optimization problem

Fig. 1.5, we propose to minimize the time of travel and to recognize that the traveler has at least two modes of locomotion, one slow (diffusion) and the other considerably faster (convection). The slow mode is placed below the smallest scale of assembly, so that every single point of the area is touched by the volumetric flow akin to diffusion. The faster channels are arranged optimally to collect the volumetric flow of each elemental building block of the network susceptible to convection. The geometric composite shape and structure formed by slow and fast flow regimes is a characteristic feature of constructal theory for volume-to-point access strategy [269].

Constructal theory gives rise to a fractal-like but deterministic structure in space and time. The "fractal" [92, 102, 109, 117] is an appropriate terminology for breaking things, which is the opposite of the direction in which natural systems evolve as predicted by constructal theory. The fractal geometry has nothing to do with time as far as descriptive geometry is concerned. The geometrical images produced by repetitive algorithms are frozen in time. The assumed algorithm can be executed, however, in both directions, from the largest to the smallest scale and vice versa. As a descriptive aid of natural phenomena, the fractal description represents a clear choice, namely from the largest to the smallest scale in an infinite number of steps. The term fractal has the concept of time built into it, that is, the act of breaking something that evolves in time from larger to smaller pieces [270]. Thus, fractals are mere descriptions but not explanations of natural phenomena [271]. There also exist alternative theories to explain pattern formation in some narrow domain of application [272, 273]. However, once again these are not universally applicable like constructal law. It can be recorded that in the manifestation of constructal law, the competition between diffusion and convection-like processes prevail.

1.3.6 Entropy Generation Minimization

Shape and structure spring from the struggle for better performance in both engineered and naturally organized systems. One good form leads to the next, as the constructal principle demands when the objective served better under global and local constraints. Each system must direct its flows to follow the thermodynamic paths that serve the global objective. To determine the most appropriate flow paths from the infinity of possible paths is the challenge of optimization of flow systems. The rate of entropy generation is a measure of how flows deviate from the ideal flow without resistances and hence without irreversibilities. Optimal flow organization minimizes entropy generation and thus maximizes the performance of systems. The method of entropy generation (EGM) or finite-time thermodynamics (FTT) is a well-established field of research [65–67, 80, 274–301].

Flows abound in nature and engineering systems are in general dissipative and thus generate entropy. Ohm's law can be employed to describe the nature of a dissipative flow as

$$R = \frac{V}{I} \tag{1.52}$$

where V is the potential that drives the current I, and R is the resistance to the flow. The flow generates entropy at the rate

$$\dot{S}_{\text{gen}} = \frac{VI}{T} \tag{1.53}$$

where T is the thermodynamic temperature at which heat transfer takes place. Using Eq. (1.53) the resistance law of Eq. (1.52) can be expressed as

$$R = \frac{T\dot{S}_{\text{gen}}}{I^2}. \tag{1.54}$$

In light of Eqs. (1.52) and (1.54), minimizing the flow resistance for a specified current I corresponds to minimizing entropy generation rate. This establishes a link between the entropy generation minimization and the constructal law.

As an example, we consider a flow tree with N branching levels. The same current I flows in each level of branching, that is, $\sum_{i=1}^{n} I_i = I$ where n is the number of passages at a certain branching level i. The flow resistance at this level of branching is

$$R = \sum_{i=1}^{n} \frac{R_i I_i}{nI}. \tag{1.55}$$

Using Eq. (1.54), the flow resistance at the same level of branching may alternatively be expressed as a quadratic average of the form

$$R = \sum_{i=1}^{n} \frac{R_i I_i^2}{I^2}. \tag{1.56}$$

Equations (1.55) and (1.56) lead to different values of resistances unless $R_i I_i = V$ is constant. Minimization of the tree resistance of Eq. (1.55) at each level under the constraint of constant current I leads to [302]

$$\sum_{i=1}^{n} \left(\frac{R_i}{nI} - \lambda_1 \right) \mathrm{d}I_i = 0 \tag{1.57}$$

where λ_1 is a constant. Similarly, minimization of the flow resistance of Eq. (1.56) under the same constraint of constant current I yields [302]

$$\sum_{i=1}^{n} \left(\frac{2R_i I_i}{I^2} - \lambda_2 \right) \mathrm{d}I_i = 0 \tag{1.58}$$

where λ_2 is another constant. In view of Eqs. (1.52)–(1.58), we may summarize

$$R = \frac{R_i}{n}, \ \frac{1}{R} = \sum_{i=1}^{n} \frac{1}{R_i}, \ I = nI_i, \ \text{and} \ V = R_i I_i. \tag{1.59}$$

These relationships obtained from the minimization of the flow tree resistance of constructal law are well known for electric currents flowing through the branched circuits. However, it is to be noted that these relations hold for any tree where the flowing current obeys the law (1.52). In Eq. (1.59), we discover the general law of equipartition of resistances. Thus, the rate of entropy generation is constant along any branching level and is expressed as

$$\dot{S}_{\text{gen}} = \sum_{i=1}^{n} \frac{R_i I_i^2}{T} = \frac{RI^2}{T}. \tag{1.60}$$

This entropy generation minimization has a special feature. The physicist Feynmann noted that [303]: "minimum principles sprang in one way or another from the least action principle of mechanics and electrodynamics. But there is a class that does not. As an example, if currents are made to go through a piece of material obeying Ohm's law, the current distribute them inside the piece so that the rate at which heat is generated is as little as possible. Also, we can say (if things are kept isothermal) that the rate at which the heat is generated is as little as possible." Thus, the least action principle accounts for point-to-point motion and cannot accommodate point-to-area or point-to-volume flows.

Hence, the constructal principle provides a wider perspective for point-to-area and point-to-volume flows. It also points out why the lowering of entropy generation gives rise to shape and structure in pursuit of minimum resistance. The generation of entropy is a consequence of the second law of thermodynamics, while the generation of flow configuration is an outcome of the constructal law. In order to meet the objective of minimum entropy generation, we have to adjust forward motivation with backward motivation.

1.3.7 Method of Intersecting Asymptotes

The geometric optimization is a direct methodology of finding optimum shape and structure suitable for a definite purpose. Let us consider the geometric optimization shown in Fig. 1.6, where a horizontal strip of high conductivity material is available for passing the current through the system, which otherwise consists of low conductivity material [304]. The volumetric ratio of high conductivity material to that of low conductivity material is r. The maximum overall resistance is between two opposite corner points where the current enters through the source and where it leaves through the sink. The optimization objective is to minimize the voltage difference between these two points required to drive the current by adjusting the shape of the system, while the respective amounts of the two components remain fixed. For an Ohmic law of current flow where the voltage drop is proportional to the current, the optimization problem is translated into an equivalent problem of maximization of current for a fixed potential drop.

When the relative volumes for the high conductivity and low conductivity material are fixed, there remains only one degree of freedom, which is the aspect ratio or slenderness ratio $s = \frac{Y}{X}$ for an actually two-dimensional problem with longitudinal dimension X, vertical dimension Y, and unit thickness in the third dimension. The upper strip has the volume $v_u = XY$ and hence the lower strip has the volume $v_l = rXY$ such that $r = \frac{v_l}{v_u}$. The resistivity ρ_l of the lower strip is smaller than the resistivity ρ_u of the upper strip. For both strips the effective

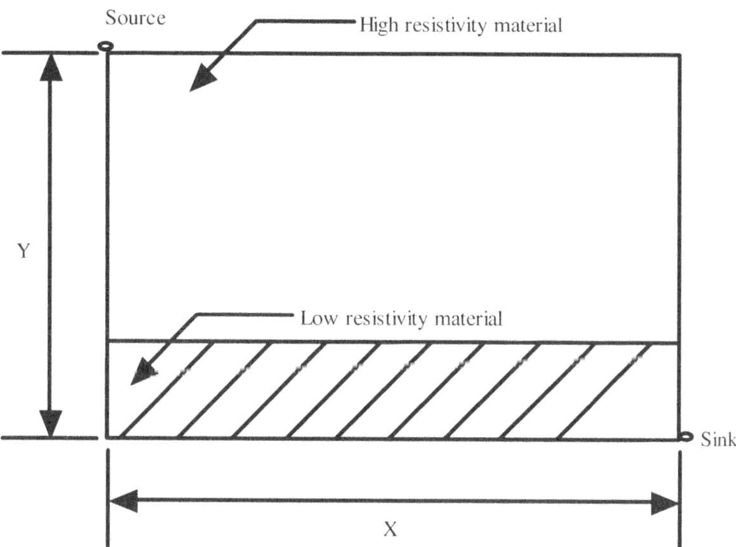

Fig. 1.6 Slender elemental volume for minimum voltage drop with fixed volumetric ratio of high to low conductivity materials

resistance is between one side and the opposite corner. The strips will have a resistance $R_s = \rho S$ where ρ is the resistivity and $S = S(X, Y)$ can be treated as a shape factor. If $X = Y = 1$ then shape factor for the strip $S_s = S(1, 1)$ will be approximately unity and will be the same for both the upper and lower strips. However, the resistance of each strip will depend on its orientation. For the lower strip where the current of interest is horizontal, the resistance falls with the cross-sectional thickness A and rises with the length in the x-direction. For the upper strip, the current of interest is vertical and similarly its resistance falls with the cross-section A and rises with the length in the y-direction.

Since the amount of the two materials is fixed, an increase in X will mean a decrease in Y maintaining the constant aspect ratio. Thus, we can assume the resistances of the strips in the following forms:

$$R_l(X, Y) \sim \rho_l S_s \frac{X}{Y} \text{ and } R_u(X, Y) \sim \rho_u S_s \frac{Y}{X} \tag{1.61}$$

where R_l and R_u are the resistances of lower and upper strip, respectively. We note that here both the regions are rectangular and thus will have a common shape and its value will not affect any internal optimization. For illustration, we assume a common square scaling shape factor of unity, which is $S_s = 1$. When $S_s \neq 1$, the effective resistivity ρ_e will be of the form $\rho_e = \rho S_s$. For $S_s = 1$, the resistances can be given as

$$R_l = \rho_l \frac{X}{A_l} = \rho_l \frac{X}{rY} \text{ and } R_u = \rho_u \frac{Y}{A_u} = \rho_u \frac{Y}{X} \tag{1.62}$$

where A_l and A_u stand for the area of the upper and lower strip, respectively.

For an Ohm's law condition, the fixed current I_0 is related to the corresponding voltage drop ΔV and resistance R by the relation $I_0 = \frac{\Delta V}{R}$.

For a common passing current, the total potential drop is the series potential drop of the individual potential drop ΔV_l of the lower strip and ΔV_u of the upper strip. Thus, we minimize total voltage drop $\Delta V = \Delta V_u + \Delta V_l$. We replace this constraint optimization problem into a free optimization employing the method of Lagrange multiplier [302]. Introducing the fixed upper volume $v = XY$, the Lagrangian can be written as

$$L = \Delta V + \lambda[v_u - XY] \tag{1.63}$$

where λ is the Lagrange multiplier. For an optimum we write

$$\frac{\partial L}{\partial X} = I_0\left[\frac{\rho_l}{rY} - \frac{\rho_u Y}{X^2}\right] - \lambda Y = 0 \tag{1.64}$$

and

$$\frac{\partial L}{\partial Y} = I_0\left[\frac{\rho_u}{X} - \frac{\rho_l}{rY^2}\right] - \lambda X = 0. \tag{1.65}$$

Eliminating the Lagrange multiplier λ yields $\frac{X^2}{Y^2} = r\frac{\rho_u}{\rho_l}$ and the optimum can be given as

$$X_{\text{opt}}^2 = v_u\sqrt{r\frac{\rho_u}{\rho_l}} \tag{1.66}$$

whence at once it follows that

$$\Delta V_l = I_0\sqrt{\frac{\rho_l \rho_u}{r}} = \Delta V_u. \tag{1.67}$$

Thus, the available potential is divided equally in the optimum arrangement where the optimum slenderness ratio is

$$s_{\text{opt}} = \left(\frac{Y}{X}\right)_{\text{opt}} = \left(\frac{\rho_l v_u}{\rho_u v_l}\right)^{1/2}. \tag{1.68}$$

Instead of using Lagrange multiplier method, the problem could also be simply solved by employing Bejan's method of intersecting asymptotes [304–310]. In the expression for potential drop $\Delta V = \Delta V_u + \Delta V_l$ we have two terms: one is proportional to the slenderness ratio $s = \frac{Y}{X}$ and the other is as its reciprocal $s^{-1} = \frac{X}{Y}$.

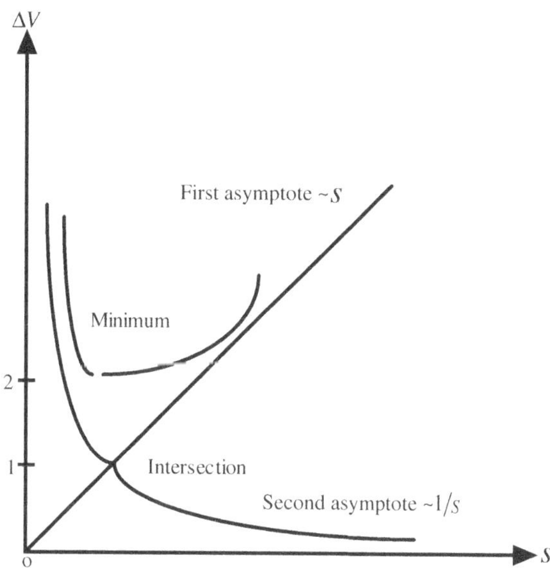

Fig. 1.7 Graphical illustration of method of intersecting asymptotes

We may readily optimize this as an ordinary differential equation in s in order to find out as before

$$\frac{\mathrm{d}}{\mathrm{d}s}(\Delta V) = \frac{\mathrm{d}}{\mathrm{d}s}\left\{I_0\left[\frac{\rho_l}{rs} + \rho_u s\right]\right\} = 0 = \rho_u - \frac{\rho_l}{rs^2} \tag{1.69}$$

such that

$$s_{\mathrm{opt}}^2 = \frac{\rho_l v_u}{\rho_u v_l}. \tag{1.70}$$

It can be seen that Eqs. (1.68) and (1.69) are but the same.

If the two separate terms are plotted against s as asymptotes as shown in Fig. 1.7, then we recognize that these two asymptotes intersect at a value of the independent argument s, which is indeed the optimum value and the true optimum value of ΔV is actually twice the optimum value of each, corresponding to equal contributions from both asymptotic terms. More generally, there may be further terms contributing to a result than the two extreme asymptotes but their intersection is likely to be a fair estimate of the true optimization-independent variable and twice the intersection value of a reasonable estimate of the optimized-dependent variable when the exact functional dependence in unknown. Thus, the method of intersecting asymptotes provides a geometrical means of visualizing the competition between the forward motivation and the backward motivation of a system.

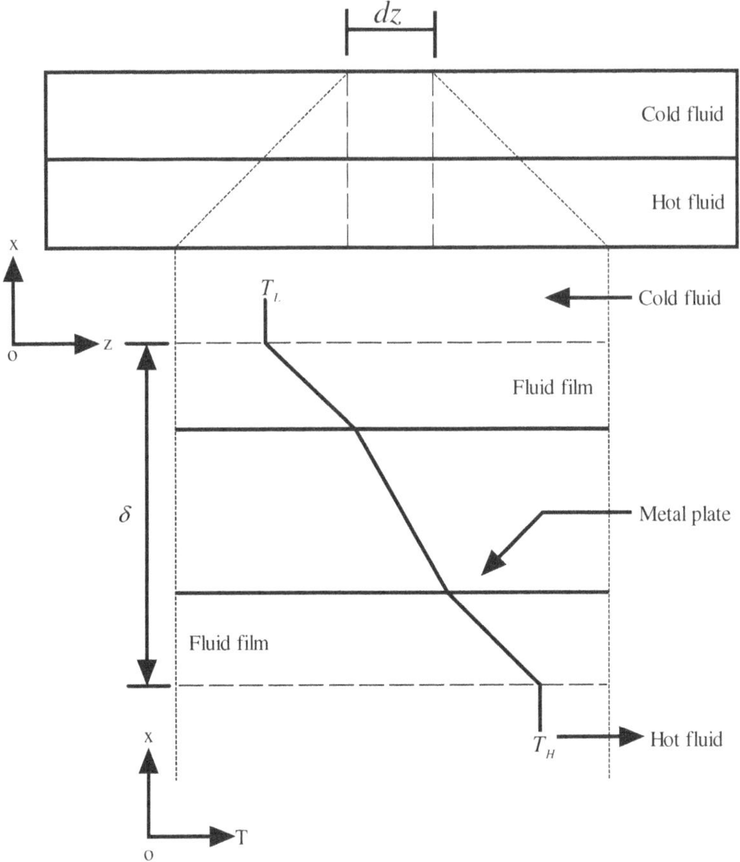

Fig. 1.8 A schematic view of the heat exchange system

1.3.8 Principle of Equipartition

In Sect. 1.3.6, we recognized in a branching network of resistors that the individual resistances, voltage drop, and the total entropy generation remain constant. In Sect. 1.3.7 the fact of equal potential drop was verified once again but without the explicit reference of entropy generation minimization. In this section we start by giving a second look at these earlier results. The following methods are but variants of entropy generation minimization principle, which is well established in engineering [65–67, 80, 274–301].

Here, we introduce the following principle in order to minimize the total entropy production for a given heat transfer rate duty $\dot{Q}$ for a heat exchanger area A. The same purpose may evolve different criteria of optimization. The heat exchange system is schematically shown in Fig. 1.8.

The cold and hot fluids are separated by a metal pipe and they flow counter currently in z-direction. Heat is exchanged only in x-direction. Assuming all variables and fluxes to be independent of y-direction, the problem is reduced to that in x–z plane. For the steady-state situation, we may write for the heat flux $J'_q(x, z)$

$$\frac{\mathrm{d}}{\mathrm{d}x}\left[J'_q(x, z)\right] = 0 \tag{1.71}$$

implying the fact that $J'_q(x, z) = J'_q(z)$. Also, $T_H = T_H(z)$ and $T_L = T_L(z)$ where T_H and T_L are the temperatures at the high temperature end and low temperature side, respectively. For a nonequilibrium but linearly irreversible heat flow process, the local entropy production rate $\dot{\sigma}(x, z)$ due to heat transfer alone is the product of heat flux $J'_q(z)$ and its conjugate driving force $X_q(z)$, i.e.,

$$\dot{\sigma}(x, z) = J'_q(z)X_q(z). \tag{1.72}$$

For heat transfer in x-direction only, the thermodynamic driving force $X_q(z)$ for entropy production is $\frac{\mathrm{d}}{\mathrm{d}x}\left(\frac{1}{T}\right)$ and thus

$$\dot{\sigma}(x, z) = J'_q(z)\frac{\mathrm{d}}{\mathrm{d}x}\left(\frac{1}{T}\right). \tag{1.73}$$

The local entropy production rate at any location z is obtained by integrating over x-direction as

$$\dot{\sigma}(z) = \int_0^\delta \dot{\sigma}(x, z)\mathrm{d}x = \int_0^\delta J'_q(z)\frac{\mathrm{d}}{\mathrm{d}x}\left(\frac{1}{T}\right)\mathrm{d}x = J'_q(z)\Delta\left[\frac{1}{T(z)}\right] \tag{1.74}$$

where δ is the thickness of the heat exchanging medium including convective films. Thus, the total entropy generation rate $\dot{S}_{\mathrm{gen}}$ of the heat exchanger is the integral of $\sigma(z)$ over the heat exchanging area A, i.e.,

$$\dot{S}_{\mathrm{gen}} = \int_A\int_\delta \sigma(z)\mathrm{d}x\mathrm{d}A = \int_A J'_q(z)\Delta\left[\frac{1}{T(z)}\right]\mathrm{d}A. \tag{1.75}$$

The heat flux may alternatively be expressed as the driving force over resistance to assume the form

$$J'_q(z) = \frac{\Delta\left[\frac{1}{T(z)}\right]}{R(z)} \tag{1.76}$$

where $R(z)$ is the resistance to the heat flow.

Then we get back the expression for entropy generation rate as

$$\dot{S}_{gen} = \int_A R(y,z) J_q'^2(y,z) \mathrm{d}y\mathrm{d}z. \tag{1.77}$$

Now, we employ Euler–Lagrange method [302] to minimize the total entropy generation rate $\dot{S}_{gen}$. For a fixed heat transfer rate duty $\dot{Q}$ we minimize, in essence, the following functional:

$$\dot{I} = \dot{S}_{gen} + \lambda\dot{Q} \tag{1.78}$$

with the constraint

$$\dot{Q} = \int_A J_q'(y,z)\mathrm{d}A \tag{1.79}$$

where λ is Lagrange multiplier. Assuming $R(y, z)$ to be independent of $J_q'(y,z)$, a differentiation yields the condition of minimum entropy production as

$$\Delta\left(\frac{1}{T}\right)_{opt} = -\frac{\lambda}{2} = \text{constant.} \tag{1.80}$$

The result contained in Eq. (1.80) is recognized as Equipartition of Forces (EoF) [311, 312] in the relevant literature.

Next, assuming symmetry in y-direction it can be seen that $R(y, z)$ or simply $R(z)$ is not independent of $J_q'(y,z)$ or truly $J_q'(z)$. First, we write

$$J_q'(z) = \frac{\Delta\left(\frac{1}{T}\right)}{R(z)} = U\Delta T(z) \tag{1.81}$$

where U is the overall heat transfer coefficient. Now, EoF principle demands that $\Delta\left(\frac{1}{T}\right)$ =constant, but again $\Delta T = \Delta T(z)$. Fourier form of heat flux is prescribed by $J_q'(z) = U\Delta T(z)$. Thus, it follows at once that $J_q'(z)$ is not a constant but a function of z. So the conjugate resistance related to the thermodynamic force is provided by

$$R(z) = \frac{\text{constant}}{J_q'(z)} = f\left(J_q'\right) = g(z) \tag{1.82}$$

where f and g are some functions. Recognizing the fact that $R(z)$ is a function of z, a better criterion for optimization can be treated as

$$\left\{J_q'(z)\Delta\left[\frac{1}{T(z)}\right]\right\}_{opt} = \text{constant} \tag{1.83}$$

or simply the local entropy production rate is represented by

$$\dot{\sigma}_{opt} = \text{constant.} \tag{1.84}$$

It follows from the result (1.84) that in order to minimize total entropy production, the local entropy production rate should be uniform throughout the passage of heat exchanger. This principle is documented as Equipartition of Entropy Production (EoEP) [313–315].

The relation between EoF and EoEP principles may be illustrated by a simple example of two time-independent parallel resistors R_1 and R_2. The forces are the voltages V_1 and V_2. For the isoforce principle we may write

$$V_1 = V_2 = V. \tag{1.85}$$

The resulting constant electric currents are $I_1 = V/R_1$ and $I_2 = V/R_2$. Thus, the supplied total current is

$$I = I_1 + I_2 = V\left(\frac{1}{R_1} + \frac{1}{R_2}\right). \tag{1.86}$$

Solving Eq. (1.86) for V we obtain the local entropy production rate as

$$\dot{\sigma}_{\mathrm{EoF}} = I_1 V_1 + I_2 V_2 = IV = \frac{I^2}{\frac{1}{R_1} + \frac{1}{R_2}}. \tag{1.87}$$

Now, if we choose to employ equipartition of entropy production rate principle, we obtain

$$\frac{\dot{\sigma}_{\mathrm{EoEP}}}{2} = I_1 V_1 = I_2 V_2 = I_1^2 R_1 = I_2^2 R_2. \tag{1.88}$$

This results in $I_1 = \sqrt{\dot{\sigma}_{\mathrm{EoEP}}/2R_1}$ and $I_2 = \sqrt{\dot{\sigma}_{\mathrm{EoEP}}/2R_2}$. The total current is thus

$$I = \sqrt{\frac{\dot{\sigma}_{\mathrm{EoEP}}}{2}}\left(\frac{1}{\sqrt{R_1}} + \frac{1}{\sqrt{R_2}}\right). \tag{1.89}$$

From Eq. (1.89) an expression for local entropy production rate is obtained as

$$\dot{\sigma}_{\mathrm{EoEP}} = \frac{2I^2}{\left(\frac{1}{\sqrt{R_1}} + \frac{1}{\sqrt{R_2}}\right)^2}. \tag{1.90}$$

Using the inequality $(1/R_1 + 1/R_2) \geq \left(1/\sqrt{R_1} + 1/\sqrt{R_2}\right)^2 / 2$ it at once follows from Eqs. (1.87) and (1.90) that

$$\dot{\sigma}_{\mathrm{EoF}} \leq \dot{\sigma}_{\mathrm{EoEP}}. \tag{1.91}$$

In Eq. (1.91), one can easily verify that the equality sign is only valid when $R_1 = R_2$. It should be noted that due to the time-independent nature of the resistors, the optimum operation is stationary. This is clearly not the case if the resistors were time-dependent. Equipartition in time implies stationary state entropy production, whereas equipartition in space means entropy production for a given position-independent force [316].

In Eq. (1.91), it is revealed that EoEP provides an upper bound of entropy production than the EoF principle. Thus, the design analysis based on EoEP accommodates a greater margin of estimate than the EoF principle. Considering the heat, flux, and temperature are to be independent of z-direction, Eq. (1.83) can be treated as

$$J_q' \Delta\left(\frac{1}{T}\right) = \text{constant}. \tag{1.92}$$

Rewriting Fourier form of heat flux as $J_q' = U\Delta T$ in Eq. (1.92) we obtain

$$U\Delta T\Delta\left(\frac{1}{T}\right) = \text{constant} \tag{1.93}$$

or

$$U(T_H - T_L)\left(\frac{1}{T_L} - \frac{1}{T_H}\right) = \text{constant}. \tag{1.94}$$

Assuming U to be fairly constant, we reach at

$$\frac{(T_H - T_L)^2}{T_H T_L} = \frac{T_H}{T_L} + \frac{T_L}{T_H} - 2 = \text{constant}. \tag{1.95}$$

Thus, we may further write

$$\tau + \frac{1}{\tau} = \text{constant} \tag{1.96}$$

where $\tau = \frac{T_H}{T_L}$. Taking the derivative with respect to z we arrive at

$$\frac{\mathrm{d}\tau}{\mathrm{d}z}\left(1 - \frac{1}{\tau^2}\right) = 0. \tag{1.97}$$

There are two solutions to this equation, viz., $|\tau| = 1$ and $\frac{\mathrm{d}\tau}{\mathrm{d}z} = 0$. For heat transport to occur at all, the solution $|\tau| = 1$ is extraneous. The solution $\frac{\mathrm{d}\tau}{\mathrm{d}z} = 0$ implies

$$\tau = \frac{T_H}{T_L} = \text{constant}. \tag{1.98}$$

Equation (1.95) can also be written in the form

$$\frac{T_H - T_L}{\sqrt{T_H T_L}} = \text{constant}. \tag{1.99}$$

With the aid of Eq. (1.98), we may write Eq. (1.99) as

$$T_H - T_L = \sqrt{\frac{1}{\text{constant}}} T_H. \tag{1.100}$$

As the resulting constant on the left side of Eq. (1.100) is much smaller than unity, we write taking the derivative with respect to z

$$\frac{\mathrm{d}}{\mathrm{d}z}(T_H - T_L) \ll \frac{\mathrm{d}T_H}{\mathrm{d}z}. \tag{1.101}$$

It follows readily from this relation that when the variations of individual temperatures of hot and cold fluid are small, we can write approximately

$$\Delta T = T_H - T_L \approx \text{constant}. \tag{1.102}$$

Equations (1.98) and (1.102) are important results for small temperature variations of hot and cold stream of fluid in a heat exchanger. Equation (1.102) as a design principle is recognized in the literature as Equipartition of Temperature Difference (EoTD) [317]. In essence, Eq. (1.102) physically represents a competition between two levels of temperatures that are in communication. Experimentally, it was found that the energy loss was minimum when the temperature difference between the hot and the cold medium was maintained constant in the liquefied natural gas heat exchangers [318].

The temperature difference between T_L and T_H can also be abridged in a process of step equilibrium. In analogy, our objective is to heat a cup of coffee while minimizing the entropy production [319]. Let there be a motel with N vacant rooms and the temperatures of each room can be set arbitrarily. The temperature of each room is adjusted to be a little higher than the previous room. By allowing the coffee cup to reach at equilibrium temperature in each room and then moving the cup to the next room at a higher temperature, we are positioned to heat the coffee cup with nominal entropy production at each step. The coffee cup is brought into the first room with temperature T_L and it equilibrates at its final value T_H in the last room. The optimization problem is to select the temperatures of the intermediate rooms in such a way that the total entropy production is minimum while the process attains complete equilibrium at each step.

At each step of equilibrium, a small amount of entropy is generated due to the heat exchange between the coffee cup and the room. The entropy generated in an infinitesimal flow of heat $\mathrm{d}Q = c\mathrm{d}T$ from the coffee cup at the temperature T to the room at constant temperature is

$$\mathrm{d}S_{\text{universe}} = \mathrm{d}S_{\text{cup}} + \mathrm{d}S_{\text{room}} = c\left(\frac{1}{T} - \frac{1}{T_i}\right)\mathrm{d}T \tag{1.103}$$

where differential change in entropy of the universe $dS_{universe}$ is contributed by that of the cup dS_{cup} and the room dS_{room} and c is the constant heat capacity of the coffee cup. Integrating to equilibrium in room i, the total change in entropy is expressed as

$$(\Delta S_{universe})_i = \int_{T_{i-1}}^{T_i} dS_{universe} = c\left(\ln\frac{T_i}{T_{i-1}} - \frac{T_i - T_{i-1}}{T_i}\right) \quad (1.104)$$

where the heat capacities of the rooms are large in comparison to that of the coffee cup. Thus, the total entropy generated throughout the N rooms is obtained upon summation to yield

$$\Delta S_{universe} = \sum_{i=1}^{N} (\Delta S_{universe})_i = c\left(\ln\frac{T_H}{T_L} + \sum_{i=1}^{N}\frac{T_{i-1}}{T_i} - N\right). \quad (1.105)$$

Once T_L and T_H are specified, the $(N - 1)$ intermediate temperatures can be chosen arbitrarily. Expanding the summation Eq. (1.105) can be written as

$$\Delta S_{universe} = c\left(\frac{T_{i-1}}{T_i} + \frac{T_i}{T_{i+1}} + \cdots\right) + c\left(\ln\frac{T_H}{T_L} - N\right). \quad (1.106)$$

The necessary condition for minimizing entropy production is stationarity with respect to a small change in each intermediate temperature, T_i for $0<i<N$. This in turn leads to

$$\frac{\partial}{\partial T_i}(\Delta S_{universe}) = c\left(-\frac{T_{i-1}}{T_i^2} + \frac{1}{T_{i+1}}\right) = 0 \quad (1.107)$$

which after simplification yields

$$\frac{T_{i-1}}{T_i} = \frac{T_i}{T_{i+1}} = \text{constant}. \quad (1.108)$$

Another principle known as Equal Thermodynamic Distance (ETD) [320] provides the same prescription for the optimal sequence of temperatures that consequentially generates minimum entropy. The thermodynamic distance, as defined below, is found as an integral over the metric in the thermodynamic variable space [321, 322] and valid for large N. The thermodynamic distance L between two states of a system is the integral of the line element

$$dL = \sqrt{-dX^T D^2 S dX} \quad (1.109)$$

where X is the column vector of extensive variables of the system and X^T is its transpose, and D^2S is the matrix of second derivatives of the entropy S with respect to the X's [323]. For a single degree of freedom, as in the case of the "coffee cup motel" problem, we have

$$\mathrm{d}L = \sqrt{-\mathrm{d}UD^2S\mathrm{d}U} = \sqrt{|D^2S|}|\mathrm{d}U| \tag{1.110}$$

where U is the internal energy of the system. Employing the relations

$$\mathrm{d}U = c\mathrm{d}T \tag{1.111}$$

and

$$\frac{\mathrm{d}^2S}{\mathrm{d}U^2} = \frac{\mathrm{d}}{\mathrm{d}U}\left(\frac{1}{T}\right) = -\frac{1}{T^2}\frac{\mathrm{d}T}{\mathrm{d}U} = -\frac{1}{cT^2} \tag{1.112}$$

we finally obtain an expression for the length element $\mathrm{d}L$ as

$$\mathrm{d}L = \sqrt{\frac{c}{T^2}(\mathrm{d}T)^2}. \tag{1.113}$$

Integrating Eq. (1.113), we obtain the thermodynamic distance $L(T_{i-1}, T_i)$ in traversing from the state of temperature T_{i-1} to T_i as

$$L(T_{i-1}, T_i) = \int_{T_{i-1}}^{T_i} \frac{\sqrt{c}}{T}\mathrm{d}T = \sqrt{c}\ln\left(\frac{T_i}{T_{i-1}}\right). \tag{1.114}$$

Since $L(T_{i-1}, T_i)$ is to be maintained constant in ETD scheme [320], the validity of Eq. (1.108) immediately follows from Eq. (1.114). The principle of ETD is also known as asymptotically optimal control process [324].

Thus, we find in the theory of macroscopic organization in nature, many different propositions of equipartition principles as variants of entropy generation minimization method, viz., equal resistance, equal potential drop, equipartition of thermodynamic forces, equipartition of local entropy production, equipartition of temperature difference, equal thermodynamic distance, etc. In simple cases, it can be examined that all such equipartition principles approximately yield the same result as provided by the rigorous method of entropy generation minimization. However, for complex problems such as temperature-dependent properties, etc., some principles are better over others in a case-specific manner, when for ease of computation a replacement of entropy generation minimization is sought. The phenomenological principle of equipartition may even appear without any specific reference to the principle of entropy generation minimization at all. In the analysis of fuel cell, equipartition of losses is observed as a consequence of optimization, which has no explicit reference to the entropy generation minimization principle [325]. In this investigation, we identify in an optimized system the macroscopic equipartition [326–330] of "entities" without specific reference to the principle of entropy generation minimization. We observe that such an equipartition principle is a characteristic feature of the law of motive force in its manifestations.

References

1. Born, M.: Natural Philosophy of Cause and Chance, p. 1. Dover, New York (1964)
2. Bejan, A.: Heat Transfer. Wiley, New York (1993)
3. Bird, R.B., Stewart, W.E., Lightfoot, E.N.: Transport Phenomena. Wiley, New York (2002)
4. Eckert, E.R.G., Drake Jr., R.M.: Analysis of Heat and Mass Transfer. McGraw-Hill, Tokyo (1972)
5. Fishenden, M., Saunders, O.A.: An Introduction to Heat Transfer. Oxford University Press, Oxford (1950)
6. Fuchs, H.U.: The Dynamics of Heat. Springer, New York (2010)
7. Ganji, D.D., Languri, E.M.: Mathematical Methods in Nonlinear Heat Transfer: A Semi-Analytical Approach. Xlibris, Indiana (2010)
8. Gröber, H., Erk, S., Grigull, U.: Fundamentals of Heat Transfer (trans: Moszyniski, J.R.). McGraw-Hill, New York (1961)
9. Isachenko, V.P., Osipova, V.A., Sukomel, A.S.: Heat Transfer (trans: Semyonov, S.). Mir, Moscow (1987)
10. Jakob, M.: Heat Transfer-I. Wiley, New York (1949)
11. Jakob, M.: Heat Transfer-II. Wiley, New York (1957)
12. Kutateladze, S.S.: Fundamentals of Heat Transfer (trans: Scripta Technica, Inc.). In: Cess, R.D. (ed.). Edward Arnold, London (1963)
13. Layton, E.T., Lienhard, J.H.: History of Heat Transfer. ASME, New York (1988)
14. Lienhard V, J.H., Lienhard IV, J.H.: A Heat Transfer Textbook. Dover, New York (2011)
15. Luikov, A.V.: Heat and Mass Transfer (trans: Kortneva, T.). Mir, Moscow (1980)
16. Lykov, A.V., Mikhailov, Y.A.: Theory of Heat and Mass Transfer (trans: Shechtman, I.). In: Hardin, R. (ed.). Israel Program for Scientific Translations, Jerusalem (1965)
17. McAdams, W.H.: Heat Transmission. McGraw-Hill, Tokyo (1958)
18. Mikhailov, M.D., Özişik, M.N.: Unified Analysis and Solutions of Heat and Mass Diffusion. Dover, New York (1984)
19. Sen, M.: Analytical Heat Transfer. University of Notre Dame, Notre Dame (2008)
20. Slattery, J.C.: Momentum, Energy, and Mass Transfer in Continua. McGraw-Hill, Tokyo (1972)
21. Weigand, B.: Analytical Methods for Heat Transfer and Fluid Flow Problems. Springer, New York (2004)
22. Batchelor, G.K.: An Introduction to Fluid Dynamics. Cambridge University Press, Cambridge (2000)
23. Brikhoff, G.: Hydrodynamics. Princeton University Press, New Jersey (1960)
24. Brodkey, R.S.: The Phenomena of Fluid Motions. Addison-Wesley, New York (1967)
25. Goldstein, S. (ed.): Modern Developments in Fluid Dynamics-I. Dover, New York (1965)
26. Goldstein, S. (ed.): Modern Developments in Fluid Dynamics-II. Dover, New York (1965)
27. Knudsen, J.G., Katz, L.D.: Fluid Dynamics and Heat Transfer. McGraw-Hill, New York (1958)
28. Landau, L.D., Lifshitz, E.M.: Fluid Mechanics (trans: Sykes, J.B., Reid, W.H.). Butterworth-Heinemann, Oxford (1987)
29. Meyer, R.E.: Introduction to Mathematical Fluid Dynamics. Dover, New York (1971)
30. Milne-Thomson, L.M.: Theoretical Hydrodynamics. Dover, New York (1968)
31. Oertel, H. (ed.): Prandtl's Essentials of Fluid Mechanics. Springer, New York (2004)

32. Shames, I.H.: Mechanics of Fluids. McGraw-Hill, New York (1992)
33. Sommerfeld, A.: Mechanics of Deformable Bodies (trans: Kuerti, G.). Academic Press, New York (1964)
34. Tokaty, G.A.: A History and Philosophy of Fluid Mechanics. Dover, New York (1971)
35. Trefil, J.S.: Introduction to the Physics of Fluids and Solids. Dover, New York (1975)
36. Whitaker, S.: Introduction to Fluid Mechanics. Prentice-Hall, New York (1968)
37. Yih, C.S.: Fluid Mechanics. McGraw-Hill, New York (1969)
38. Bejan, A.: Advanced Engineering Thermodynamics. Wiley, New York (2006)
39. Bridgman, P.W.: The Nature of Thermodynamics. Harvard University Press, Cambridge (1943)
40. Buchdahl, H.A.: The Concepts of Classical Thermodynamics. Cambridge University Press, Cambridge (1966)
41. Callen, H.B.: Thermodynamics and an Introduction to Thermostatistics. Wiley, New York (2005)
42. Clausius, R.: The Mechanical Theory of Heat (trans: Browne, W.R.). Macmillan, London (1879)
43. Epstein, P.S.: Textbook of Thermodynamics. Wiley, New York (1937)
44. Fermi, E.: Thermodynamics. Dover, New York (1956)
45. Giles, R.: Mathematical Foundations of Thermodynamics. Pergamon, Oxford (1964)
46. Gyftopoulos, E.P., Beretta, G.P.: Thermodynamics: Foundations and Applications. Dover, New York (2005)
47. Hatsopoulos, G.N., Keenan, J.H.: Principles of Generalized Thermodynamics. Wiley, New York (1965)
48. Hoffman, E.J.: Analytic Thermodynamics. Taylor & Francis, New York (1991)
49. Hsieh, J.S.: Principles of Thermodynamics. McGraw-Hill, Tokyo (1975)
50. Kestin, J.: A Course in Thermodynamics-I. Blaisdell, Waltham (1966)
51. Kestin, J.: A Course in Thermodynamics-II. McGraw-Hill, New York (1979)
52. Kirillin, V.A., Sychev, V.V., Sheindlin, A.E.: Engineering Thermodynamics (trans: Semyonov, S.). Mir, Moscow (1981)
53. Landsberg, P.T.: Thermodynamics and Statistical Mechanics. Oxford University Press, Oxford (1978)
54. Lavenda, B.H.: A New Perspective of Thermodynamics. Springer, New York (2010)
55. Müller, I.: A History of Thermodynamics. Springer, New York (2007)
56. Pauli, W.: Thermodynamics and the Kinetic Theory of Gases. Dover, New York (1973)
57. Pippard, A.B.: The Elements of Classical Thermodynamics. Cambridge University Press, Cambridge (1957)
58. Planck, M.: Treatise on Thermodynamics (trans: Ogg, A.). Dover, New York (1945)
59. Sommerfeld, A.: Thermodynamics and Statistical Mechanics (trans: Kestin, J.). In: Bopp, F., Meixner, J. (eds.). Academic Press, New York (1964)
60. Sychev, V.V.: The Differential Equations of Thermodynamics (trans: Yankovsky, E.). Mir, Moscow (1983)
61. Thess, A.: The Entropy Principle. Springer, New York (2011)
62. Tisza, L.: Generalized Thermodynamics. MIT Press, Cambridge (1966)
63. Truesdell, C.: Rational Thermodynamics. Springer, New York (1984)
64. Truesdell, C., Bharatha, S.: The Concepts and Logic of Classical Thermodynamics as a Theory of Heat Engines. Springer, New York (1977)
65. Bejan, A.: Advanced Engineering Thermodynamics, pp. 101–144, 574–655. Wiley, New York (2006)
66. Bejan, A.: Entropy Generation Through Heat and Fluid Flow. Wiley, New York (1982)
67. Bejan, A.: Thermodynamic optimization of inanimate and animate flow systems. In: Bejan, A., Mamut, E. (eds.) Thermodynamic Optimization of Complex Energy Systems. Kluwer, Dordrecht (1999)
68. Thompson, B.: An enquiry concerning the source of heat which is excited by friction. Philos. Trans. R. Soc. Lond. **88**, 80–102 (1798)

69. Truesdell, C.: Six Lectures on Modern Natural Philosophy, pp. 100–101. Springer, New York (1966)
70. Bejan, A.: Shape and Structure, from Engineering to Nature, p. xvi. Cambridge University Press, Cambridge (2000)
71. Rankine, W.J.M.: A Manual of the Steam Engine and Other Prime Movers, p. xv. In: Miller, W.J. (ed.). Charles Griffin & Co., London (1888)
72. Munk, M.M.: My early aerodynamic research—thoughts and memories. Ann. Rev. Fluid Mech. **13**, 1–7 (1981)
73. Ribenboim, P.: Fermat's Last Theorem for Amateurs. Springer, New York (1999)
74. Schlichting, H., Gersten, K.: Boundary-Layer Theory (trans: Mayes, K.), pp. 145–164. Springer, New York (2000)
75. Franklin, A.: Principle of inertia in the middle ages. Am. J. Phys. **44**, 529–543 (1976)
76. Mario, B.: Mach's critique of Newtonian mechanics. Am. J. Phys. **34**, 585–596 (1966)
77. Sciama, D.W.: On the origin of inertia. Mon. Not. R. Astron. Soc. **113**, 34–42 (1953)
78. Ghosh, A.: Origin of Inertia: Extended Mach Principle and Cosmological Consequences. Aperion, Canada (2000)
79. Barbour, J., Pfister, H. (eds.): Mach's Principle: From Newton's Bucket to Quantum Gravity, p. 530. Brikhauser, Boston (1995)
80. Bejan, A.: Second-law analysis in heat transfer and thermal design. Adv. Heat Transf. **15**, 1–58 (1982)
81. Pramanick, A.K.: Philosophy of nature, Reflection magazine, pp. 2–5. R.E. College, Durgapur (1990–1991)
82. Mendoza, E. (ed.): Reflections on the Motive Power of Fire. Dover, New York (2005)
83. Carnot, S.: Reflections on the motive power of fire, and on machines fitted to develop that power (trans: Thurston, R.H. (ed.)). In: Mendoza, E. (ed.) Reflections on the Motive Power of Fire. Dover, New York (2005)
84. Dijksterhuis, E.J.: Archimedes (trans: Dikshoorn, C.), pp. 16–17. Princeton University Press, Princeton (1987)
85. Galilei, G.: Dialogues Concerning Two New Sciences (trans: Crew, H., De Salvio, A.), pp. 62–65. Dover, New York (1914)
86. Kant, I.: Metaphysical Foundations of Natural Science. In: Friedman, M. (ed.). Cambridge University Press, Cambridge (2004)
87. Meyer, K. (ed.): Ørstad's Works-III, pp. 151–190. Andr. Fred. Høst, Copenhagen (1920)
88. Newton, I.: Mathematical Principles of Natural Philosophy and His System of the World (trans: Motte, A., Cajori, F.), pp. 16–17. University of California Press, Berkeley (1946)
89. Sorensen, R.A.: Thought Experiments. Oxford University Press, Oxford (1992)
90. Fermi, E.: Thermodynamics, pp. 94–97. Dover, New York (1956)
91. Glasstone, S.: Thermodynamics for Chemists, pp. 273–316, 462–500. Van Nostrand, New York (1947)
92. Avnir, D., Biham, O., Lidar, D., Malcai, O.: Is the geometry of nature fractal? Science **279**, 39–40 (1998)
93. Cook, T.A.: The Curves of Life. Dover, New York (1979)
94. Cohn, D.L.: Optimal systems-I: vascularized system. Bull. Math. Biophys. **16**, 59–74 (1954)
95. French, M.: Invention and Evolution: Design in Nature and Engineering. Cambridge University Press, Cambridge (1994)
96. Haldane, J.B.S.: On Being the Right Size. Oxford University Press, Oxford (1928)
97. Kepler, J.: The Six-Cornered Snowflake (trans: Hardie, C. (ed.)). Oxford University Press, Oxford (1976)
98. Kuramoto, Y.: Chemical Oscillations, Waves, and Turbulence. Dover, New York (2009)
99. Lawton, J.H.: More time means more variations. Nature **334**, 563 (1988)
100. Leopold, L.B.: Rivers. Am. Sci. **50**, 511–537 (1962)
101. MacDonald, N.: Trees and Networks in Biological Models. Wiley, Chichester (1983)
102. Mandelbrot, B.B.: The Fractal Geometry of Nature. Freeman, New York (1983)
103. McMahon, T.A.: Size and shape in biology. Science **179**, 1201–1204 (1973)

104. Meinhardt, H.: Models of Biological Pattern Formation. Academic Press, London (1982)
105. Morisawa, M.: Rivers: Form and Process. Longman, London (1985)
106. Müller-Krumbhaar, H.: Formation, competition and selection of growth patterns. In: Puri, S., Dasgupta, S. (eds.) Nonlinearities in Complex Systems, pp. 121–135. Narosa, New Delhi (1997)
107. Ottino, J.M.: The mixing of fluids. Sci. Am. **260**, 40–49 (1989)
108. Pauling, L.: Apparent icosahedral symmetry is due to directed multiple twining of cube crystals. Nature **317**, 512–514 (1985)
109. Peitgen, H.-O., Richter, P.H.: The Beauty of Fractals. Springer, New York (1986)
110. Penrose, R.: The role of aesthetics in pure and applied mathematical research. Bull. Inst. Math. Appl. **10**, 266–271 (1974)
111. Prigogine, I.: From Being to Becoming. Freeman, New York (1980)
112. Rouse, H., Ince, S.: History of Hydraulics. Dover, New York (1963)
113. Salem, J., Wolfram, S.: Thermodynamics and hydrodynamics of cellular automata. In: Wolfram, S. (ed.) Theory and Applications of Cellular Automata, pp. 362–366. World Scientific, Philadelphia (1986)
114. Schattschneider, D.: Visions of Symmetry. Freeman, New York (1991)
115. Schmidt-Nielsen, K.: Scaling: Why is Animal Size so Important? Cambridge University Press, Cambridge (1984)
116. Schuster, H.G.: Deterministic Chaos. Physik-Verlag, Weinheim (1986)
117. Scroeder, M.: Fractals, Chaos, Power Laws. Dover, New York (2009)
118. Shannon, C.E.: Prediction and entropy of printed english. Bell Syst. Tech. J. **30**, 50–64 (1951)
119. Stewart, I., Golubitsky, M.: Fearful Symmetry. Dover, New York (2011)
120. Thompson, D.W.: On Growth and Form. Cambridge University Press, Cambridge (1942)
121. Uman, M.A.: Lightning. Dover, New York (1984)
122. Vogel, S.: Life's Devices. Princeton University Press, Princeton (1988)
123. West, G.B., Brown, J.H., Enquist, B.J.: A general model for the origin of allometric scaling laws in biology. Science **276**, 122–126 (1997)
124. Weyl, H.: Symmetry. Princeton University Press, Princeton (1989)
125. Wilson, T.A.: Design of the bronchial tree. Nature **213**, 668–669 (1967)
126. Zimmerman, M.H.: Hydraulic architecture of some diffuse-porous trees. Can. J. Bot. **56**, 2286–2295 (1978)
127. Bejan, A.: Advanced Engineering Thermodynamics, pp. 63–64. Wiley, New York (2006)
128. Jammer, M.: Concepts of Mass in Classical and Modern Physics. Dover, New York (1997)
129. Jammer, M.: Concepts of Space. Dover, New York (2012)
130. Reichenbach, H.: The Direction of Time. In: Reichenbach, M. (ed.). Dover, New York (1999)
131. Weyl, H.: Space Time Matter (trans: Brose, H.L.). Dover, New York (1952)
132. Alekseev, G.N.: Energy and Entropy (trans: Taube, U.M.). Mir, Moscow (1986)
133. Mach, E.: History and Root of the Principle of Conservation of Energy (trans: Jourdain, P.E.B.). Open Court, Chicago (1911)
134. Müller, I., Weiss, W.: Entropy and Energy. Springer, New York (2005)
135. Glansdorff, P., Prigogine, I.: Thermodynamic Theory of Structure, Stability and Fluctuations. Wiley-Interscience, New York (1971)
136. Bunge, M.: Casuality: The Place of Causal Principle in Modern Science. World Publishing, New York (1963)
137. Bridgman, P.W.: The Nature of Thermodynamics, pp. 3–115. Harvard University Press, Cambridge (1943)
138. Bridgman, P.W.: The Nature of Thermodynamics, pp. 116–179. Harvard University Press, Cambridge (1943)
139. Abraham, R.H., Shaw, C.D.: Dynamics—The Geometry of Behaviour. Addison-Wesley, Cambridge (1992)
140. Appell, P.: Traité de Mécanique Rationnelle-I. Gauthier-Villars, Paris (1941) (in French)

141. Appell, P.: Traité de Mécanique Rationnelle-II. Gauthier-Villars, Paris (1953) (in French)
142. Appell, P.: Traité de Mécanique Rationnelle-III. Gauthier-Villars, Paris (1952) (in French)
143. Appell, P.: Traité de Mécanique Rationnelle-IV. Gauthier-Villars, Paris (1932) (in French)
144. Appell, P.: Traité de Mécanique Rationnelle-V. Gauthier-Villars, Paris (1926) (in French)
145. Arnol'd, V.I.: Mathematical Methods of Classical Mechanics (trans: Vogtmann, K., Weinstein, A.). Springer, New York (1995)
146. Bernoulli I, J., Bernoulli II, N.: Die Werke von Johann I und Nicolaus II Bernoulli. Band 6. Mechanik-I. Birkhäuser, Boston (2008) (in German)
147. Chandrasekhar, S.: Newton's Principia for the Common Reader. Oxford University Press, Oxford (1995)
148. Chetaev, N.G.: Theoretical Mechanics (trans: Aleksanova, I., Starzhinsky, V.M. (ed.)). In: Rumyantsev, V.V., Yakimova, K.E. (eds.). Mir, Moscow (1989)
149. D' Le Alembert, J.R.: Traité de Dynamanique. David, Paris (1743) (in French)
150. Dugas, R.A.: History of Mechanics (trans: Maddox, J.R.). Dover, New York (2011)
151. Fung, Y.C.: First Course in Continuum Mechanics. Prentice Hall, New Jersey (1993)
152. Galilei, G.: On Mechanics (trans: Drake, S.). University of Wiscosin, Madison (1960)
153. Goldstein, H.: Classical Mechanics. Addison-Wesley, Cambridge (1980)
154. Hertz, H.: Principles of Mechanics Presented in a New Form. Dover, New York (2003)
155. Lagrange, J.L.: Analytical Mechanics (trans: Boissonnade, A., Vagliente, V.N. (ed.)). Kulwer, Massachusetts (2010)
156. Landau, L.D., Lifshitz, E.M.: Mechanics (trans: Sykes, J.B., Bell, J.S.). Butterworth-Heinemann, Oxford (2010)
157. Mach, E.: The Science of Mechanics (trans: McCormack, T.J.). Open Court, Chicago (1942)
158. Newton, I.: Philosophiae Naturalis Principia Mathematica (trans: Motte, A., Cajori, F., Revised). University of California, Berkeley (1934)
159. Osgood, W.F.: Mechanics. Macmillan, New York (1937)
160. Parkaus, H.: Mechanik der Festen Körper. Springer, Viena (1959) (in German)
161. Planck, M.: General Mechanics (trans: Brose, H.L.). Macmillan, London (1933)
162. Poisson, S.D.: A Treatise of Mechanics (trans: Harte, H.H.). Longman, London (1842)
163. Sommerfeld, A.: Mechanics (trans: Stern, M.O.). Academic Press, New York (1964)
164. Spencer, A.J.M.: Continuum Mechanics. Dover, New York (2004)
165. Synge, J.L., Griffith, B.A.: Principles of Mechanics. McGraw-Hill, New York (1959)
166. Truesdell, C.: The Elements of Continuum Mechanics. Springer, New York (1984)
167. Bigelow, J., Ellis, B., Pargetter, R.: Forces. Philos. Sci. **55**, 614–630 (1988)
168. Cohen, I.B.: Newton's second law and the concept of force in the Principia. In: Palter, R. (ed.) The Annus Mirabilis of Sir Isaac Newton 1666–1966, pp. 143–185. MIT Press, Cambridge (1970)
169. Ellis, B.D.: The existence of force. Stud. Hist. Philos. Sci. **7**, 171–185 (1976)
170. Erlichson, H.: Motive force and centripetal force in Newton's mechanics. Am. J. Phys. **59**, 842–849 (1991)
171. Jammer, M.: Concepts of Force. Dover, New York (1999)
172. Pourcaiu, B.: Newton's interpretation of Newton's second law. Arch. Hist. Exact Sci. **60**, 157–207 (2006)
173. Pourciau, B.: Is Newton's second law really Newton's? Am. J. Phys. **79**, 1015–1022 (2011)
174. Steinner, A.: The story of force: from Aristotle to Einstein. Phys. Educ. **29**, 77–85 (1996)
175. Westfall, R.S.: Force in Newton's Physics. Macdonald, London (1971)
176. Feynman, R.P., Leighton, R.B., Sands, M.: The Feynman Lectures on Physics-I, pp. 107–109. Narosa, New Delhi (2003)
177. Lanczos, C.: The Variational Principles of Mechanics, pp. 88–110. Dover, New York (1986)
178. Sommerfeld, A.: Mechanics (trans: Stern, M.O.), pp. 48–58. Academic Press, New York (1964)
179. Love, A.E.H.: A Treatise on the Mathematical Theory of Elasticity, pp. 171–172. Dover, New York (2011)

180. Courant, R., Hilbert, D.: Methods of Mathematical Physics-II, pp. 32–39. Wiley-VCH, Berlin (2008)
181. Sokolnikoff, I.S.: Mathematical Theory of Elasticity, pp. 387–389. McGraw-Hill, New York (1956)
182. Langhaar, H.L.: Energy Methods in Applied Mechanics, pp. 126–130. Wiley, New York (1962)
183. Sommerfeld, A.: Mechanics (trans: Stern, M.O.), pp. 181–185. Academic Press, New York (1964)
184. Landau, L.D., Lifshitz, E.M.: Mechanics (trans: Sykes, J.B., Bell, J.S.), pp. 1–12. Butterworth-Heinemann, Oxford (2010)
185. Sachs, M.: Ideas of the Theory of Relativity, pp. 101–103. Israel University Press, Jerusalem (1947)
186. Landau, L.D., Lifshitz, E.M.: The Classical Theory of Fields (trans: Hamermesh, M.). Butterworth-Heinemann, Oxford (1980)
187. Morse, P.M., Feshbach, H.: Methods of Theoretical Physics-I, pp. 1–118, 200–222, 275–347, 1759–1901. McGraw-Hill, New York (1953)
188. Maxwell, J.C.: A Treatise on Electricity and Magnetism-I. Dover, New York (1954)
189. Maxwell, J.C.: A Treatise on Electricity and Magnetism-II. Dover, New York (1954)
190. Morse, P.M., Feshbach, H.: Methods of Theoretical Physics-I, pp. 21–31. McGraw-Hill, New York (1953)
191. Feynman, R.P., Leighton, R.B., Sands, M.: The Feynman Lectures on Physics-II, pp. 685–688. Narosa, New Delhi (2001)
192. Shames, I.H.: Mechanics of Fluids, pp. 49–51. McGraw-Hill, New York (1992)
193. Feynman, R.P., Leighton, R.B., Sands, M.: The Feynman Lectures on Physics-II, pp. 702–703. Narosa, New Delhi (2001)
194. Crank, J.: The Mathematics of Diffusion. Oxford University Press, Oxford (1980)
195. Glasstone, S., Laidler, K.J., Eyring, H.: The Theory of Rate Processes. McGraw-Hill, New York (1941)
196. Jacobs, M.H.: Diffusion Processes. Springer, New York (1967)
197. Fourier, J.B.J.: Mémoire d'analyse sur le mouvement de la chaleur dans les fluides. In: Mémoires de l'Académie Royale des Sciences de l'Institut de France, pp. 507–530. Didot, Paris (1833) (Presented 4 Sept 1820) (in French)
198. Fourier, J.B.J.: Oeuvres de Fourier-II, pp. 595–614. In: Darboux, G. (ed.). Gauthier-Villars, Paris (1890) (in French)
199. Poisson, S.D.. Théorie Mathématique de la Chaleur, p. 86. Bachelier, Imprimeur-Libraire, Paris (1835) (in French)
200. Bubnov, V.A.: On generalized hydrodynamic equations used in heat transfer theory. Int. J. Heat Mass Transf. **16**, 109–119 (1973)
201. Kasterin, N.P.: Generalization of Aerodynamic and Electrodynamic Fundamental Equations. Indz. Akad. Nauk, Moscow (1937) (in Russian)
202. Patankar, S.V.: Numerical Heat Transfer and Fluid Flow, pp. 15–17. Taylor & Francis, New York (2011)
203. Akhiezer, N.I.: The Calculus of Variations. Blaisdell, New York (1962)
204. Aleksandrov, A.D., Kolmogorov, A.N., Lavrent'ev, M.A. (eds.): Mathematics-III (trans: Gould, S.H. (ed.)), pp. 119–138. Dover, New York (1999)
205. Becker, M.: The Principles and Applications of Variational Methods. MIT Press, Cambridge (1964)
206. Biot, M.A.: Variational Principles in Heat Transfer. Oxford University Press, Oxford (1970)
207. Bliss, G.A.: Calculus of Variations. Open Court, Chicago (1925)
208. Bolza, O.: Lectures on the Calculus of Variations. Chelsea, New York (1973)
209. Courant, R., Hilbert, D.: Methods of Mathematical Physics-I, pp. 164–274. Wiley-VCH, New York (2004)
210. Donnelly, R.J., Herman, R., Prigogine, I. (eds.): Non-Equilibrium Thermodynamics, Variational Techniques and Stability. University of Chicago Press, Chicago (1966)

211. Dreyfus, S.E.: Dynamic Programming and the Calculus of Variations. Academic Press, New York (1965)
212. Elsǵolc, L.D.: Calculus of Variations. Dover, New York (2007)
213. Forsythe, A.R.: Calculus of Variations. Cambridge University Press, Cambridge (1927)
214. Gelfand, I.M., Fomin, S.V.: Calculus of Variations (trans: Silverman, R.A.). Dover, New York (2000)
215. Goldstein, H.H.: A History of the Calculus of Variations from the 17th Through the 19th Century. Springer, New York (1980)
216. Hestenes, M.R.: Calculus of Variations and Optimal Control Theory. Wiley, New York (1966)
217. Hildebrand, F.B.: Methods of Applied Mathematics, pp. 119–221. Dover, New York (1992)
218. Kumar, I.J.: Recent mathematical methods in heat transfer. Adv. Heat Transf. **8**, 22–33 (1972)
219. Leitmann, G.: Calculus of Variations and Optimal Control. Plenum, New York (1981)
220. Lemons, D.S.: Perfect Form. Princeton University Press, Princeton (1997)
221. Mikhlin, S.G.: Variational Methods in Mathematical Physics (trans: Boddington, T., Chambers, L.I.G. (ed.)). Macmillan, New York (1964)
222. Moiseiwitsch, B.L.: Variational Principles. Dover, New York (2004)
223. Morse, P.M., Fesbach, H.: Methods of Theoretical Physics-II, pp. 1106–1172. McGraw-Hill, New York (1953)
224. Nesbet, R.K.: Variational Principle and Methods in Theoretical Physics and Chemistry. Cambridge University Press, Cambridge (2005)
225. Petrov, I.P.: Variational Methods in Optimal Control Theory. Academic Press, New York (1968)
226. Pinch, E.R.: Optimal Control and the Calculus of Variations. Oxford University Press, Oxford (1993)
227. Rektorys, K. (ed.): Survey of Applicable Mathematics (trans: Výborný, R., et al.), pp. 1020–1044. Iliffe Books, London (1969)
228. Schechter, R.S.: The Variational Method in Engineering. McGraw-Hill, New York (1967)
229. Sieniutycz, S.: Conservation Laws in Variational Thermo-Hydrodynamics. Springer, New York (1994)
230. Weinstock, R.: Calculus of Variations. Dover, New York (1974)
231. Yong, L.C.: Lectures on the Calculus of Variations and Optimal Control Theory. Chelsea, New York (1980)
232. Yourgrau, W., Mandelstam, S.: Variational Principles in Dynamics and Quantum Theory. Dover, New York (2007)
233. Bliss, G.A.: Lectures on the Calculus of Variations, pp. 10–12. Chicago University Press, Chicago (1959)
234. Hildebrand, F.B.: Methods of Applied Mathematics, pp. 139–145. Dover, New York (1992)
235. Pontryagin, L.S., Boltyanskii, V.G., Gamkrelidze, R.V., Mishchenko, E.F.: The Mathematical Theory of Optimal Processes (trans: Trirogoff, K.N.). In: Neustadt, L.W. (ed.). Wiley-Interscience, New York (1965)
236. Feynman, R.P., Leighton, R.B., Sands, M.: The Feynman Lectures on Physics-I, pp. 315–323. Narosa, New Delhi (2003)
237. Sommerfeld, A.: Optics (trans: Laporte, O., Moldauer, P.A.), pp. 355–356. Academic Press, New York (1964)
238. Weinstock, R.: Calculus of Variations, pp. 67–71. Dover, New York (1974)
239. Sommerfeld, A.: Optics (trans: Laporte, O., Moldauer, P.A.), pp. 150–152. Academic Press, New York (1964)
240. Bejan, A.: Advanced Engineering Thermodynamics, pp. 705–841. Wiley, New York (2006)
241. Bejan, A.: Constructal tree network for fluid flow between a finite-size volume and one source or sink. Int. J. Therm. Sci. **36**, 592–604 (1997)
242. Bejan, A.: Shape and Structure, from Engineering to Nature. Cambridge University Press, Cambridge (2000)

243. Bejan, A.: Street network theory of organization in nature. J. Adv. Transp. **30**, 85–107 (1996)
244. Bejan, A.: Theory of organization in nature. Int. J. Heat Mass Transf. **40**, 2097–2104 (1997)
245. Bejan, A., Dincer, I., Lorente, S., Miguel, A.F., Reis, A.H.: Porous and Complex Flow Structures in Modern Technologies. Springer, New York (2004)
246. Bejan, A., Lorente, S.: Design with Constructal Theory. Wiley, New York (2008)
247. Bejan, A., Lorente, S.: La loi constructale (trans: Kremer-Marietti, A.). L'Harmattan, Paris (2005)
248. Bejan, A., Lorente, S., Miguel, A.F., Reis, A.H. (eds.): Along with Constructal Theory. Faculty of the Geosciences and the Environment, University of Lausanne Press, Lausanne (2006)
249. Bejan, A., Zane, J.P.: Design in Nature: How the Constructal Law Governs Evolution in Biology, Physics, Technology, and Social Organization. Anchor Books, New York (2013)
250. Poirier, H.: Une Théorie Explique L'intelligence de la Nature. Sci. Vie. **1034**, 44–63 (2003) (in French)
251. Reis, A.H.: Constructal theory: from engineering to physics, and how flow systems develop shape and structure. Appl. Mech. Rev. **59**, 269–282 (2006)
252. Rocha, L.A.O., Lorente, S., Bejan, A. (eds.): Constructal Law and the Unifying Principle of Design. Springer, New York (2013)
253. Rosa, R.N., Reis, A.H., Miguel, A.F. (eds.): Bejan's Constructal Theory of Shape and Structure. Évora Geophysics Center, University of Évora, Portugal (2004)
254. Bejan, A.: Advanced Engineering Thermodynamics, p. 807. Wiley, New York (1997)
255. Bejan, A.: Constructal-theory network for conducting paths for cooling a heat generating volume. Int. J. Heat Mass Transf. **40**, 799–816 (1997)
256. Bejan, A.: Advanced Engineering Thermodynamics, pp. 705–707. Wiley, New York (1997)
257. Bern, M.W., Graham, R.L.: The shortest network problem. Sci. Am. **260**, 84–89 (1989)
258. Courant, R., Robbins, H.: What is Mathematics? (Stewart, I., Revised), pp. 354–361. Oxford University Press, Oxford (2007)
259. Gilbert, E.N., Pollak, H.O.: Steiner minimal trees. SIAM J. Appl. Math. **16**, 1–29 (1968)
260. Gray, A.: Tubes. Addision-Wesley, New York (1990)
261. Hwang, F.K., Richards, D., Winter, P.: The Steiner's Tree Problem. Elsevier, New York (1992)
262. Ivanov, A.O., Tuzhilin, A.A.: Branching Solution to One-Dimensional Variational Problems. World Scientific, Philadelphia (2001)
263. Ivanov, A.O., Tuzhilin, A.A.: Minimal Networks: The Steiner Problem and its Generalizations. CRC Press, New York (1994)
264. Lyusternik, L.A.: Shortest Paths Variational Problems (trans: Brown, R.B., Collins, P.). Macmillan, New York (1964)
265. Newman, M., Barabási, A.L., Watts, D.J. (eds.): The Structure and Dynamics of Networks. Princeton University Press, Princeton (2006)
266. Rubinstein, J.H., Thomas, D.A.: A variational approach to the Steiner network problem. Ann. Oper. Res. **33**, 481–499 (1991)
267. Stewart, I.: Trees, telephones, and tiles. New Sci. **1795**, 26–29 (1991)
268. Winter, P.: Steiner problems in networks: a survey. Networks **17**, 129–187 (1987)
269. Bejan, A.: Advanced Engineering Thermodynamics, p. 808. Wiley, New York (1997)
270. Bejan, A.: Advanced Engineering Thermodynamics, pp. 741–742. Wiley, New York (1997)
271. Bloom, A.L.: Geomorphology, p. 204. Prentice-Hall, New Jersey (1978)
272. Davis, S.H.: Theory of Solidification. Cambridge University Press, Cambridge (2001)
273. Xu, J.J.: Dynamical Theory of Dendritic Growth in Convection Flow. Kulwer, Boston (2010)
274. Ahern, J.E.: The Exergy Method of Energy Systems Analysis. Wiley, New York (1980)
275. Alefeld, G.: Die Exergie und der II. Hauptsatz der Thermodynamik. Brennstoff-Wärme-Kraft **40**, 458–464 (1988) (in German)

276. Andresen, B.: Finite-Time Thermodynamics. Physics Laboratory II, University of Copenhagen, Copenhagen (1983)
277. Andresen, B., Salamon, P., Berry, R.S.: Thermodynamics in Finite-Time. Phys. Today **37**, 62–70 (1984)
278. Bejan, A.: A study of entropy generation in fundamental convective heat transfer. J. Heat Transf. **101**, 718–725 (1979)
279. Bejan, A.: Engineering advances on finite-time thermodynamics. Am. J. Phys. **62**, 11–12 (1994)
280. Bejan, A.: Entropy generation minimization: the new thermodynamics of finite-size devices and finite-time processes. J. Appl. Phys. **79**, 1191–1218 (1996)
281. Bejan, A.: Notes on the history of the method of entropy generation minimization (finite time thermodynamics). J. Non-Equilib. Thermodyn. **21**, 239–242 (1996)
282. Bejan, A., Tsatsaronis, G., Moran, M.: Thermal Design and Optimization. Wiley, New York (1996)
283. Berry, R.S., Kazakov, V., Sieniutycz, S., Szwast, Z., Tsirlin, A.M.: Thermodynamic Optimization of Finite-Time Processes. Wiley, Chichester (1999)
284. Bošnjaković, F.: Thechnische Thermodynamik-I. Steinkopf, Dresden (1953) (in German)
285. Brodyanskii, V.M.: Exergy Method of Thermodynamic Analysis. Energia, Moscow (1973) (in Russian)
286. Bruges, E.A.: Available Energy and the Second Law Analysis. Butterworth, London (1959)
287. Chambadal, P.: Evolution et Applications du Concept d'Entropie, Sec. 30. Dunod, Paris (1963) (in French)
288. Chen, L., Sun, F. (eds.): Advances in Finite Time Thermodynamics: Analysis and Optimization. Nova Science, New York (2004)
289. De Vos, A.: Endoreversible Thermodynamics of Solar Energy Conversion. Oxford University Press, Oxford (1992)
290. Gouy, G.: Sur l'energie utilizable. J. Phys. **8**, 501–518 (1889) (in French)
291. Kestin, J.: Availability: the concept and associated terminology. Energy **5**, 679–692 (1980)
292. Kolenda, Z.S., Donizak, J., Hubert, J.: On the minimum entropy production in steady state heat conduction process. In: Proceedings of Efficiency, Cost, Optimization, Simulation and Environmental Impact of Energy Systems (ECOS'02), Berlin, 3–5 July 2002
293. Mironova, V.A., Tsirlin, A.M., Kazakov, V.A., Berry, R.S.: Finite-time thermodynamics: exergy and optimization of time-constrained processes. J. Appl. Phys. **76**, 629–636 (1994)
294. Moran, M.J.: Availability Analysis: A Guide to Efficient Energy Usage. ASME, New York (1989)
295. Moran, M.J., Sciubba, E. (eds.): Second Law Analysis of Thermal Systems. ASME, New York (1987)
296. Nerescu, I., Radcenco, V.: Exergy Analysis of Thermal Processes. Editura Technica, Bucharest (1970) (in Romanian)
297. Novikov, I.I.: Thermodynamics. Mashinostroenie, Moscow (1984) (in Russian)
298. Sieniutycz, S., Salamon, P. (eds.): Finite-Time Thermodynamics and Thermoeconomics. Taylor & Francis, New York (1990)
299. Stodola, A.: Steam and Gas Turbines. McGraw-Hill, New York (1910)
300. Szargut, J.: International progress in second law analysis. Energy Int. J. **5**, 709–718 (1980)
301. Tolman, R.C., Fine, P.C.: On the irreversible production of entropy. Rev. Mod. Phys. **20**, 51–77 (1948)
302. Lewins, J.D.: Introducing the Lagrange multiplier to engineering mathematics. Int. J. Mech. Eng. Educ. **22**, 191–207 (1994)
303. Feynman, R.P., Leighton, R.B., Sands, M.: The Feynman Lectures on Physics-II, pp. 956–974. Narosa, New Delhi (2001)
304. Bejan, A.: Shape and Structure, from Engineering to Nature, pp. 29–41, 45–49, 163–174. Cambridge University Press, Cambridge (2000)
305. Bejan, A.: Convection Heat Transfer, pp. 136–141, 211–214, 225–228, 279–286, 404–406, 591–593, 613–616. Wiley, New York (2004)

306. Bejan, A., Dincer, I., Lorente, S., Miguel, A.F., Reis, A.H.: Porous and Complex Flow Structures in Modern Technologies, pp. 58–66, 201–212. Springer, New York (2004)
307. Bejan, A., Lorente, S.: Design with Constructal Theory, pp. 81–96, 364–349. Wiley, New York (2008)
308. Lewins, J.: Bejan's constructal theory of equal potential distribution. Int. J. Heat Mass Transf. **46**, 1541–1543 (2003)
309. Nield, D.A., Bejan, A.: Convection in Porous Media, pp. 275–282. Springer, New York (2006)
310. Sadeghipour, M.S., Razi, Y.P.: Natural convection from a confined horizontal cylinder: the optimal distance between the confining walls. Int. J. Heat Mass Transf. **44**, 367–374 (2001)
311. Nummedal, L., Kjelstrup, S.: Equipartition of forces as a lower bound on the entropy production in heat exchange. Int. J. Heat Mass Transf. **44**, 2827–2833 (2001)
312. Sauar, E.: Energy efficient process design by equipartition of forces. Doctoral thesis, Norwegian University of Science and Technology, Trondheim (1998)
313. Johannessen, E., Nummedal, L., Kjelstrup, S.: Minimizing the entropy production in heat exchange. Int. J. Heat Mass Transf. **45**, 2649–2654 (2002)
314. Nummedal, L.: Entropy production minimization of chemical reactors and heat exchangers. Doctoral thesis, Norwegian University of Science and Technology, Trondheim (1998)
315. Tondeur, D., Kvaalen, E.: Equipartition of entropy production: an optimality criterion for transfer and separation processes. Ind. Eng. Chem. Res. **26**, 50–56 (1987)
316. Bedeaux, D., Standaert, F., Hemmes, K., Kjelstrup, S.: Optimization of processes by equipartition. J. Non-Equilib. Thermodyn. **24**, 242–259 (1999)
317. Balkan, F.: Comparison of entropy minimization principles in heat exchange and a short-cut principle: EoTD. Int. J. Energy Res. **27**, 1003–1014 (2003)
318. Fredheim, A.: Thermal design of coil-wound LNG heat exchangers. Shell-side heat transfer and pressure drop. Doctoral thesis, Norwegian Institute of Technology, Trondheim (1994)
319. Salamon, P., Nulton, J.D., Siragusa, G., Limon, A., Bedeaux, D., Kjelstrup, S.: A simple example of control to minimize entropy production. J. Non-Equilib. Thermodyn. **27**, 45–55 (2002)
320. Nulton, J., Salamon, P., Andresen, B., Anmin, Q.: Quasistatic processes as step equilibrations. J. Chem. Phys. **83**, 334–338 (1985)
321. Wielhold, F.: Geometric representation of equilibrium thermodynamics. Acc. Chem. Res. **9**, 236–240 (1976)
322. Wienhold, F.: Thermodynamics and geometry. Phys. Today **29**, 23–30 (1976)
323. Salamon, P., Nulton, J.D.. The geometry of separation process: the horse-carrot theorem for steady flow systems. Europhys. Lett. **42**, 571–576 (1998)
324. Salamon, P., Nulton, J.D., Siragusa, G., Andresen, T.R., Limon, A.: Principles of control thermodynamics. Energy **26**, 307–319 (2001)
325. Standaert, F.: Analytical fuel cell modelling and exergy analysis of fuel cells. Doctoral thesis, Technical University of Delft, Delft (1998)
326. Bejan, A.: Advanced Engineering Thermodynamics, pp. 352–356, 464–466, 569–571, 709–721, 782–788, 816–820. Wiley, New York (2006)
327. Bejan, A.: Shape and Structure, from Engineering to Nature, pp. 53–56, 84–88, 99–108, 151–161, 220–223, 234–242, 287–288. Cambridge University Press, Cambridge (2000)
328. Bejan, A., Tondeur, D.: Equipartition, optimal allocation, and the constructal approach to predicting organization in nature. Rev. Gen. Therm. **37**, 165–180 (1998)
329. De Vos, A., Desoete, B.: Equipartition principle in finite-time thermodynamics. J. Non-Equilib. Thermodyn. **25**, 1–13 (2000)
330. Pramanick, A.K., Das, P.K.: Note on constructal theory of organization in nature. Int. J. Heat Mass Transf. **48**, 1974–1981 (2005)

Chapter 2
Conductive Heat Transport Systems

Thermodynamics gives me two strong impressions: first of a subject not yet complete or at least one of whose ultimate possibilities have not yet been explored, so that perhaps there may still be further generalizations awaiting discovery; and secondly and even more strongly as a subject whose fundamental and elementary operations have never been subject to adequate analysis.

P. W. Bridgman

In this chapter, we directly apply the law of motive force in place of variational formulation as well as optimal control theory for a class of problems pertaining to conductive heat transport mode in the realm of thermal insulation design. From the physics of the principle it has been deduced that a truly minimum exists for such class of problems. To start with, the optimum distribution of limited amount of insulating material on one side of a plane surface as well as a curved wall is obtained assuming that the amount of insulating material does not affect the imposed temperature gradient. Next, we apply the same physical theory for a more general case when a stream of fluid is suspended in a different temperature, and where the volume of insulation material does affect the temperature distribution. Finally, it has been argued that Schmidt's criterion for the fin design, tangent law of conductive heat transport and the Fermat's principle in geometrical optics are but special stipulations of the proposed law of nature, whereas the constructal law is a stand-alone principle where the proposed law of motive force is manifested through the competition of backward and forward motivation of slower (diffusion-like) and faster (convection-like) processes.

2.1 The Problem

The problem of optimization is the very essence of reality [1–4]. It is well known that many physical theories naturally give rise to the variational optimization principle from which the governing equations of the system can be deduced; the class of theories that do not yield a spontaneous variational formulation on account of nonlinearity, or else can be modified to admit a variational form [5]. Inversely, it also follows at once that the laws of physical theories when expressed as differential equations, the possibility of their reduction to a variational principle is evident from purely mathematical reasoning and does not depend on certain

A. K. Pramanick, *The Nature of Motive Force*, Heat and Mass Transfer,
DOI: 10.1007/978-3-642-54471-2_2, © Springer-Verlag Berlin Heidelberg 2014

attributes intrinsic to the theory [6]. Despite these mathematical assertions, remarkably the classical thermodynamics [7–9] usually formulated is devoid of variational principles. However, it can be shown that as far as the implications for quasistatic transitions are concerned, the second law of thermodynamics can be formulated as a variational principle [10]. In classical mechanics, it can be established that by means of Gauss's principle [11], all problems may be reduced to those pertaining to maxima and minima and, hence possibly, to a problem of variational calculus. Thus, the variational technique as an optimization procedure has undergone tremendous upsurge both in science and engineering [12–19]. However, physicists and engineers often seem to disagree about the meaning of a variational principle [20]. For physicists, the fundamental element is generally the existence of a Lagrangian function through which the governing equations of the system are obtained by taking the functional derivatives. The main appeal of the Lagrange function is its power of synthesis. The whole physics of the problem is expressed in terms of a single function. But the Lagrangian in our extended sense exists only for dissipative systems. On the other hand, for engineers the main point often seems to be the existence of a variational technique, as clearly indicated by the type of approximation methods [21] employed in engineering optimization, which are largely independent of the existence of a Lagrange function. Variational principle can also be formulated [22] outside the postulate of minimum entropy production [23] and the concept of local potential [23]. Quite apart from variational formulation, a wide class of practical optimization problems can be expressed in the form of the Pontryagin maximum principle [24]. It is reported that attempts to solve these problems by the method of classical calculus of variations are not attractive [25].

An optimization procedure, such as variational method, is usually carried out halfway, that is, the values of the parameters of a trial function are found for which a property of the system under consideration, such as the energy, reaches its optimum value [26]. Thus, the current research methodology emphasizes the physical understanding of the problem in thermodynamic optimization of systems with particular examples in mind.

The present contribution explores the proposed law of motive force, a physical principle, mainly for the design of conductive insulation systems, which was recently analyzed by the formal method of calculus of variations [27]. Thermodynamic optimization of insulation system is also historically important and remains an active research frontier in the contemporary heat transfer research. It is historically important because the chronology of the entropy generation minimization field [28] began with the design of insulation systems subject to finite-size constraint [29]. It is an active area of research since power plant and refrigeration unit can be regarded as thermal insulation system [30], while accepting the general definition that thermal insulation is a system that prevents two surfaces of different temperatures from coming into direct thermal communication. From the physical perspective of the problem, it is demonstrated that for such a class of optimization problems a truly minimum exists. Finally, it has been argued that there should be some basis for analogies among physical theories [31]. Being persuaded by such basis the manifestation of law of motive force in Fermat's principle [32, 33] and constructal law [34, 35] from which geometric forms [36] can be deduced out of a single physics principle is sought. The current contribution examines the result obtained by the author [37] in the light of the proposed law of motive force.

2.2 A Physical Principle in Heat Transport

To engineer nature is to understand her first. In this endeavor we seek continually a more general principle than the existing till an all-encompassing theory is established. The speculative way of seeking a new law is but to guess it first [38]. In a nutshell, the law of motive force enunciates to identify the "conservation" of some physical quantities as a physical principle of thermodynamic optimization. Existence of such "isolines" is one of the most fundamental characteristics of extremality. Guided by this line of thought we proceed to identify the contributing competing mechanisms that constitute the locus of the physical process path describing the isoline.

To illustrate the rudimental feature of this principle, we first consider a plane wall of length L and width W perpendicular to the plane of the paper as shown in Fig. 2.1. The wall temperature variation $T(x)$ is only along the longitudinal direction x. The fundamental question corners around how to distribute a finite amount of insulating material either with constant or varying thickness $t(x)$ on the wall for minimum heat loss.

The insulated wall can be thought of being pieced into m equal or unequal length of sections. The more the local distribution of unit insulation material ΔV, the less the local heat transfer rate Δq in general. On the other hand, making a particular segment of the wall more effective leads other parts of the wall to be less effective in insulation. Thus, we identify heat transfer and insulation volume to be two competing physical factors (forces, motives) in insulation design. Here, the incidence of heat transfer acts as a forward motivation, whereas insulation volume plays the role of backward motivation. Hence, following the proposition of law of motive force, the legitimate postulate should be the uniform (equal) effectiveness of the insulation. This natural law translates mathematically into

$$\Delta q_i + \lambda_i \Delta V = \Delta q + \lambda \Delta V = C_{sv} \tag{2.1a}$$

for $i = 1, 2, 3, \ldots, m$ and where C_{sv} is a constant. We drop the subscript i for equal segmentation. Here, λ is a numerical and dimensional factor which makes the volume, a physical quantity, to be dimensionally homogeneous with another physical entity heat. The far reaching consequences of this parameter in a greater perspective are to be realized [39]. The order of magnitude of the parameter λ is such that for which the problem of optimization is nontrivial. Hypothetically, there may be some portion of the wall not covered with insulation at all, meaning $\lambda = 0$ as in the leading and trailing edges of the wall. On the contrary, all insulation

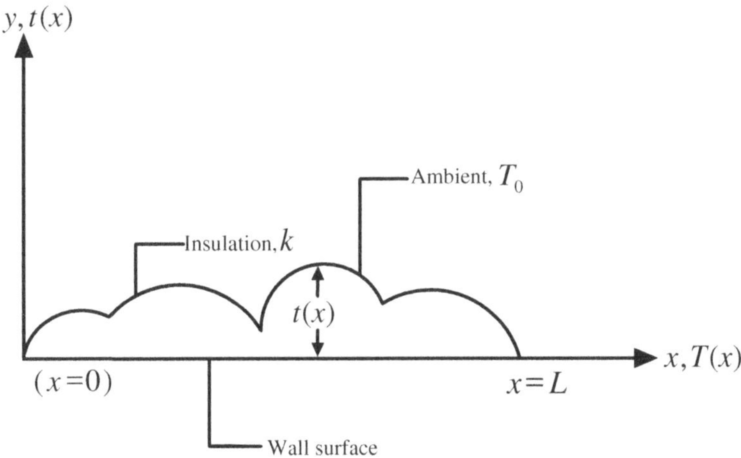

Fig. 2.1 A flat plate with arbitrary variation in insulation thickness

material can be applied onto a limited spot, leading to $\lambda > 0$. Thus, from the physical point of view the dimensional scale factor λ is bounded only in the domain $[0, \infty]$. To realize in another way the role played by λ, Eq. (2.1a) may be written in an alternative fashion when one of the constituents leads to a constant as

$$\frac{\Delta q_i}{\lambda_i \Delta V} = \frac{\Delta q}{\lambda \Delta V} = C_{rv} \tag{2.1b}$$

for $i = 1, 2, 3, \ldots, m$ and where C_{rv} is another constant. Notationally, the subscript is dropped for equal segmentations as before. It can be seen that for $\lambda \to 0$, the constant on the right side of Eq. (2.1b) tends to a very high value, meaning a very high rate of heat transfer as also indicated by Eq. (2.1a) and thus not a desirable feature for modeling. On the other hand, for $\lambda \to \infty$ the constant on the right side of Eq. (2.1b) runs to a very low value implying a very low heat transfer as also implied by Eq. (2.1a), and thus ensures a favorable modeling feature. But at the same time for the optimization problem to be nontrivial, the material volume cannot be unlimited or scarce posing a restriction to the upper and lower bounds for the value of λ too.

Either Eq. (2.1a) or Eq. (2.1b) can be employed, as the case may be for the ease of computation or applicability, to obtain optimal profile of insulation in connection with minimum heat transfer from the wall with definite curvature and temperature profile.

If we consider insulating a line element instead of a plane wall, Eqs. (2.1a) and (2.1b) transform, respectively, into

$$\Delta q_i + \mu_i \Delta A = \Delta q + \lambda \Delta A = C_{sa} \tag{2.2a}$$

and

$$\frac{\Delta q_i}{\mu_i \Delta A} = \frac{\Delta q}{\mu \Delta A} = C_{ra} \tag{2.2b}$$

where the volume element ΔV is replaced by the surface area element ΔA and the dimensional role of μ has been changed to that of λ.

It is interesting to report that Eq. (2.2b) resembles that of Schmidt's idea [40] of optimum profile shape for cooling fin with minimum weight. At the same time, it is to be noted that Schmidt's criterion was obtained on a different heuristic logic. The intuitive logic of Schmidt was confirmed through rigorous variational formulation by Duffin [41]. Jany and Bejan [42] came to the conclusion that the idea of fin shape optimization has an important analog in the design of long ducts for fluid flow.

It is thought-provoking that the problem for maximum heat transfer objective resembles the challenge of insulation design for minimum heat transfer. It truly reflects the opposing action of the motive forces [43] as forward and backward motivation in apparently two antagonistic arrangements. The physical factor that transcribes a problem of insulation into a question of fin is the curvature of the surface in consideration. For example, critical insulation thickness [44] exists only in reality for the design of cylindrical and spherical layers, but not in the sizing of plane or nearly plane layers. Thus, we repeat the symmetric appearance of a physical principle [45–47] with respect to its foundation in mathematical terms [31].

2.3 The Physical Basis for Extremum Heat Transfer

The criteria for distinguishing between the maximum and minimum values of the functional have been investigated by many eminent mathematicians [48]. A rigorous mathematical discussion of the discriminating conditions may be found from the fundamental principle alone [49]. In our present endeavor we will however, provide a physical basis for the existence of the extremum. To be specific with the domain of application of this analysis, we take the example of purely conductive insulation system.

From the physical perspective, heat transfer and insulation volume are both nonnegative quantities. It is to be noted that we did not adopt here a control volume approach so as to regard heat transfer as positive or negative with respect to the system in a conventional manner. Again, Eq. (2.1a) truly represents a competition between two opposing tendencies of the system: backward motivation and forward motivation. Further, their constancy of summation leads to the fact that increment of one quantity drives to the decrement of the other in numerical estimate. These logics translate into the following mathematical prescriptions

$$\Delta q = \Delta Q^2,\ \Delta V = \Delta v^2,\ \text{and}\ \lambda = -\psi. \tag{2.3}$$

Thus, Eq. (2.1a) transforms into

$$\Delta Q^2 - \psi \Delta v^2 = C_{sv}. \tag{2.4}$$

Since we are interested in global extremum, integrating upon Eq. (2.4) the entire length of the plate, we have

$$\int_0^L (\Delta Q^2 - \psi \Delta v^2)\mathrm{d}x = C_{sv} L. \tag{2.5}$$

As indicated by the first example of the use of trigonometric series in the theory of heat [50], we adopt Fourier expansion [51] for the pattern of distribution of insulating material in primitive variables to be

$$\Delta v = \sum_{m=1}^{\infty} a_m \sin \frac{m\pi}{L} x \tag{2.6}$$

where a_m's are some constants compatible with the convergence of the series. Rearranging Eq. (2.1a) in the following form:

$$\Delta q = C_{sv} - \lambda \Delta V \tag{2.7}$$

and recognizing that heat transfer takes place in a normal direction to the plane under consideration [52], we find a compatible [53, 54] Fourier series as

$$\Delta Q = \sum_{m=1}^{\infty} \frac{m\pi}{L} a_m \cos \frac{m\pi}{L} x. \tag{2.8}$$

Invoking Parseval's theorem [55] to the relations (2.6) and (2.8) we arrive, respectively, at

$$\int_0^L \Delta v^2 \mathrm{d}x = \frac{L}{2} \sum_{m=1}^{\infty} a_m^2 \tag{2.9}$$

and

$$\int_0^L \Delta Q^2 \mathrm{d}x = \frac{L}{2} \sum_{m=1}^{\infty} \frac{m^2 \pi^2}{L^2} a_m^2. \tag{2.10}$$

The mathematical prescription for the applicability of Parseval's theorem is that

$$\Delta v(0) = \Delta v(L) = 0 \tag{2.11}$$

and ΔQ whose square is Lebesgue integrable [56] over the interval [0, L]. From the physics of the problem these criteria are quite recognizable. Thus, incorporating Eqs. (2.9) and (2.10) into Eq. (2.5) we have

$$C_{sv} = \frac{1}{2}\sum_{m=1}^{\infty}\left(\frac{m^2\pi^2}{L^2} - \psi\right)a_m^2. \tag{2.12}$$

Noting that the left-hand side is a finite positive quantity and hence there exists a minimum for the parameter ψ in the range

$$\psi < \frac{\pi^2}{L^2}. \tag{2.13}$$

It is to be remarked that the physical role [57] played by the parameter ψ here is different from λ in Eq. (2.1a). The physical argument presented above easily extends to the Sturm–Liouville theory [58]. Thus, we conclude that a truly minimum exists for this class of problems of insulation design. Next, we will calculate only the optimum profile for different geometries and temperature distributions. Once thus obtained optimum profile tallies with the established results, the minimum heat transfer quantity follows at once.

2.4 Temperature Distribution and Heat Transfer from an Insulated Wall

In many engineering applications [27], a nonlinear temperature variation $T(x)$ in the longitudinal direction x of the wall of finite length L arises with definite curvature $\mathrm{d}^2T/\mathrm{d}^2x$. When the curvature of the wall temperature function is positive, temperature profile of the wall can be outlined as

$$\frac{T(x) - T_0}{T_L - T_0} = \frac{\exp\left(n\frac{x}{L}\right) - 1}{e^n - 1} \tag{2.14a}$$

where T_0 and T_L are wall temperatures at $x = 0$ and $x = L$, respectively, and the nondimensional parameter n bears the same sign as the curvature of the wall temperature function. Here, T_0 is also the ambient temperature. For the curvature of the temperature function of the wall to be negative, temperature distribution of the wall can be expressed algebraically as

$$\frac{T(x) - T_0}{T_L - T_0} = \frac{1 - \exp\left(n\frac{x}{L}\right)}{1 - e^n}. \tag{2.14b}$$

In case of vanishingly small curvature of temperature function, passing to the limit $n \to 0$ either from Eq. (2.14a) or Eq. (2.14b), we obtain by applying L'Hospital's theorem a linear temperature distribution as

$$\frac{T(x) - T_0}{T_L - T_0} = \frac{x}{L}. \tag{2.14c}$$

On the other hand, the mathematical advantage of the exponential representation of temperature is that it can be readily treated in resulting differential equations [59]. Hence, it may be possible to cast the whole exercise as a control problem of differential equation alone [60].

Recognizing the local temperature gradient $\Delta T = T(x) - T_0$ to be the cause of spontaneous heat transfer effect Δq in a coupled conductive–convective formulation the expression for heat transfer stands as

$$\Delta q = \frac{\Delta T}{\frac{F[t(x)]}{k\Delta A} + \frac{1}{h\Delta A}} \tag{2.15a}$$

where ΔA is the elemental heat transferring area, k is the constant thermal conductivity of the insulating material, h is the local convective heat transfer coefficient, $t(x)$ is the local thickness of insulation, and $F[t(x)]$ is the function of insulation thickness. Passing to the limit $h \to \infty$ in Eq. (2.15a) we arrive at

$$\lim_{h\to\infty} \Delta q = \lim_{h\to\infty} \frac{\Delta T}{\frac{F[t(x)]}{k\Delta A} + \frac{1}{h\Delta A}} = \frac{\Delta T}{\frac{F[t(x)]}{k\Delta A}}. \tag{2.15b}$$

Equation (2.15a) is connected to Eq. (2.15b) in the same manner as the two-dimensional problem of heat transfer is related to the one-dimension when either of the dimensions is very great in comparison with the other.

In mathematical modeling of the problem we have both the choices: either to consider or not the effect of local insulation thickness on the driving force (ΔT) for the heat transfer.

2.5 Insulation on Plane Surface with Static Wall Temperature Condition

By static wall temperature condition we mean that the temperature distribution on the wall will not be affected by the amount of insulation mounted. We consider here a plane wall of length L and width W. The average insulation thickness $\bar{t}$ can be defined on total volume V as

$$\bar{t} = \frac{V}{WL} = \frac{1}{L}\int_0^L t(x)\mathrm{d}x. \tag{2.16}$$

In Eq. (2.15b) we recognize for a plane wall that

$$\frac{F[t(x)]}{\Delta A} = \frac{t(x)}{W\mathrm{d}x}. \tag{2.17}$$

Now, employing the law of motive force (2.1b) in Eq. (2.15b) along with Eq. (2.17) we directly obtain

$$t(x) = \left(\frac{\lambda}{k}\right)^{1/2} (\Delta T)^{1/2}. \tag{2.18}$$

For linear temperature distribution, invoking Eq. (2.14c) in Eq. (2.18) we arrive at

$$t(x) = \Lambda_1 \left(\frac{x}{L}\right)^{1/2} \tag{2.19}$$

where Λ_1 is the shorthand for the constant $\left[\frac{\lambda}{k}(T_L - T_0)\right]^{1/2}$. Integrating Eq. (2.19) between 0 and L and employing Eq. (2.16) for the definition of average thickness we have

$$\Lambda_1 = \frac{3}{2}\bar{t}. \tag{2.20}$$

Optimal insulation profile is obtained by eliminating the constant Λ_1 between Eqs. (2.19) and (2.20) as

$$t_{1l}(x) = \frac{3}{2}\bar{t}\left(\frac{x}{L}\right)^{1/2}. \tag{2.21}$$

When the curvature of the wall temperature function is positive, employing Eq. (2.14a) in Eq. (2.18) and adopting similar procedure, we get optimal insulation thickness as function

$$t_{2l}(x) = \frac{n}{2}\bar{t}\frac{\sqrt{\exp\left(n\frac{x}{L}\right) - 1}}{\sqrt{e^n - 1} - \tan^{-1}\sqrt{e^n - 1}}. \tag{2.22}$$

For the curvature of the wall temperature profile to be negative, recruiting Eq. (2.14b) to Eq. (2.18), similarly we obtain the optimal insulation profile as

$$t_{3l}(x) = \frac{n}{2}\bar{t}\frac{\sqrt{1 - \exp\left(n\frac{x}{L}\right)}}{\sqrt{1 - e^n} - \tanh^{-1}\sqrt{1 - e^n}}. \tag{2.23}$$

2.6 Insulation on Cylindrical Surface with Static Wall Temperature Condition

As stated before, static wall temperature condition implies that the temperature of the wall is not a function of insulation volume. We now consider a cylinder of radius r and length L. Geometrically, we mean a situation with the surface of revolution of the plane wall mounted with arbitrary insulation volume along with a translation in the vertical direction. Such a description bears easy extension to the fundamental problem presented in Fig. 2.1. Then the fixed volume V of insulation is rendered by

$$V = \int_0^L \pi r^2 \left\{ \left[1 + \frac{t(x)}{r} \right]^2 - 1 \right\} \mathrm{d}x. \tag{2.24a}$$

The relative thickness of insulation material is obtained in dimensionless form as

$$\bar{V} = \frac{V}{\pi r^2 L} = \frac{1}{L} \int_0^L \left\{ \left[1 + \frac{t(x)}{r} \right]^2 - 1 \right\} \mathrm{d}x. \tag{2.24b}$$

When the wall thickness is not negligibly relative to the radius of the curvature of the wall surface, the problem must be analyzed by a method that takes the curvature into account. In Eq. (2.15b) we identify for a cylindrical wall [61]

$$\frac{F[t(x)]}{\Delta A} = \frac{\ln\left[1 + \frac{t(x)}{r}\right]}{2\pi \mathrm{d}x}. \tag{2.25}$$

Plugging Eq. (2.25) in lieu of Eq. (2.15b) along with Eq. (2.24a) into the law of motive force (2.1a), we get the optimal insulation profile to comply with the following condition:

$$\delta \ln \delta = \left(\frac{k}{\lambda r^2} \right)^{1/2} (\Delta T)^{1/2} \tag{2.26a}$$

where

$$\delta(x) = 1 + \frac{t(x)}{r}. \tag{2.26b}$$

Assuming a linear temperature distribution (2.14c) in Eq. (2.26a) we obtain

$$\delta \ln \delta = \Lambda_2 \left(\frac{x}{L} \right)^{1/2} \tag{2.27a}$$

where Λ_2 is the notation for the parameter $\left[\frac{k}{\lambda r^2}(T_L - T_0)\right]^{1/2}$. The constant Λ_2 is determined from the definition (2.24b) as

$$\Lambda_2 = \left[\frac{2}{\bar{V}} \int_1^{\Delta} \delta(\delta^2 - 1) \ln \delta \ln(e\delta) \mathrm{d}\delta\right]^{1/2} \tag{2.27b}$$

where

$$\Delta = 1 + \frac{t_{\mathrm{opt}}(L)}{r}. \tag{2.27c}$$

Eliminating the constant Λ_2 between Eqs. (2.27a) and (2.27b) optimal insulation profile is obtained as

$$\delta \ln \delta = \left[\frac{2}{\bar{V}} \int_1^{\Delta} \delta(\delta^2 - 1) \ln \delta \ln(e\delta) \mathrm{d}\delta\right]^{1/2} \left(\frac{x}{L}\right)^{1/2}. \tag{2.28}$$

In the event of positive wall temperature curvature recruiting Eq. (2.14a) in Eq. (2.26a) we have

$$\delta \ln \delta = \Lambda_3 \left[\exp\left(n\frac{x}{L}\right) - 1\right]^{1/2} \tag{2.29a}$$

where Λ_3 is the shorthand for the group $\left[\frac{k}{\lambda r^2} \frac{T_L - T_0}{e^n - 1}\right]^{1/2}$. The constant Λ_3 is implicitly determined using definition (2.24b) as

$$\frac{2}{n} \int_1^{\Delta} \frac{\delta(\delta^2 - 1) \ln \delta \ln(e\delta)}{(\delta \ln \delta)^2 + \Lambda_3} \mathrm{d}\delta = \bar{V}. \tag{2.29b}$$

Eliminating the constant term Λ_3 between Eqs. (2.29a) and (2.29b) we obtain the required optimum insulation profile.

Similarly, for negative curvature of the wall temperature function employing Eq. (2.14b) in Eq. (2.26a) and exercising the same procedure, we obtain the optimal profile of insulation as the eliminant of the parametric constant Λ_4 between the following equations:

$$\delta \ln \delta = \Lambda_4 \left[1 - \exp\left(n\frac{x}{L}\right)\right]^{1/2} \tag{2.30a}$$

and

$$\frac{2}{n} \int_1^{\Delta} \frac{\delta(\delta^2 - 1) \ln \delta \ln(e\delta)}{(\delta \ln \delta)^2 - \Lambda_4} \mathrm{d}\delta = \bar{V} \tag{2.30b}$$

where Λ_4 is the shorthand for the constant $\left[\frac{k}{\lambda r^2} \frac{T_L - T_0}{1 - e^n}\right]^{1/2}$.

2.7 Insulation on Cylindrical Surface with Dynamic Wall Temperature Condition

Unlike in Sects. 2.5 and 2.6, we consider here a dynamic local temperature gradient situation for the wall. In other words, we do not neglect the effect of local insulation thickness on the local temperature distribution. Rather, we impose the more realistic condition that the local temperature distribution is affected by the amount of insulation. Now, as a modeling feature we are at liberty to apply insulation in such a way that the local temperature potential remains piecewise constant, that is, $\Delta T \neq \Delta T(x)$. This makes in turn the local overall heat transfer coefficient U to be independent of longitudinal spatial position [62], that is, again $U \neq U(x)$. This idea of equipartitioned (uniformed) potential difference is due to the author [63].

Let us consider a stream of fluid with local temperature distribution $T_f(x)$ passing through an insulated cylindrical tube whose outer surface is exposed to a constant environment temperature T_0 such that

$$\Delta T = T_f(x) - T_0 = C_T \tag{2.31}$$

where C_T is a constant.

The expression for overall heat transfer coefficient U between the local bulk temperature of the stream $T_f(x)$ and the environment at T_0 can be readily obtained from any standard heat transfer textbook [61] as

$$\frac{1}{U2\pi r\mathrm{d}x} = \frac{1}{h_0 2\pi[r + t(x)]\mathrm{d}x} + \frac{\ln\left[1 + \frac{t(x)}{r}\right]}{k_i 2\pi\mathrm{d}x} + \frac{t_w}{k_w 2\pi r\mathrm{d}x} + \frac{1}{h_i 2\pi r\mathrm{d}x} \tag{2.32a}$$

where h_i and h_0 are the local convective heat transfer coefficients for the inner fluid and the outer fluid, k_i and k_w are the conductivities of the insulating material and cylinder wall, respectively, r is the inner radius of the wall, t_w is the thickness of the wall. Recognizing the fact that $h_i, h_0 \to \infty$ and $\frac{t_w}{r} \to 0$, we pass on to these limits in Eq. (2.32a) to obtain

$$\frac{1}{U2\pi r\mathrm{d}x} = \frac{\ln\left[1 + \frac{t(x)}{r}\right]}{k_i 2\pi\mathrm{d}x}. \tag{2.32b}$$

Putting Eq. (2.32b) into Eq. (2.25) along with Eq. (2.15b) into the law of motive force (2.1a) we obtain

$$\left[\frac{\Delta T}{U2\pi r} + \lambda\pi r^2\left\{\left[1 + \frac{t(x)}{r}\right]^2 - 1\right\}\right]\mathrm{d}x = \text{constant.} \tag{2.33a}$$

By definition ΔT and U are constants and since $\mathrm{d}x$ can be arbitrarily small, the bracketed quantity on the right side vanishes identically, i.e.,

$$\frac{\Delta T}{U2\pi r} + \lambda\pi r^2\left\{\left[1 + \frac{t(x)}{r}\right]^2 - 1\right\} = 0. \tag{2.33b}$$

Since $t(x)$ is the only variable on the left side, the physical solution of the equation leads to the fact that

$$t(x) = \text{constant}. \tag{2.34}$$

The constant of the right side of Eq. (2.34) is determined from Eq. (2.24b) as

$$t_{\text{opt}} = r\left[(1 + \bar{V})^{1/2} - 1\right]. \tag{2.35}$$

Equation (2.35) is an important result and was obtained using the calculus of variations [27] and optimal control theory [64] as reported in literature as well as traditionally practiced by engineers.

2.8 Law of Motive Force, Tangent Law, Fermat's Principle, and Constructal Law

For two different materials of the wall and the insulating volume to be in perfect thermal contact, the interfacial boundary conditions demand that [65]

$$-k_1\left(\frac{\partial T_1}{\partial y}\right)_{0^+} = -k_2\left(\frac{\partial T_2}{\partial y}\right)_{0^-} \tag{2.36a}$$

and

$$T_1 = T_2 \tag{2.36b}$$

where the subscripts 1 and 2 refer to the general wall and the insulating material, respectively. Equation (2.36a) can be written as

$$\frac{\left(\frac{\partial T_1}{\partial y}\right)_{0^+}}{\left(\frac{\partial T_2}{\partial y}\right)_{0^-}} = \frac{k_2}{k_1} \tag{2.37a}$$

which readily admits the following form:

$$\frac{\tan\theta_1}{\tan\theta_2} = \frac{k_2}{k_1} \tag{2.37b}$$

where θ_1 and θ_2 are the angles of incidence and refraction, respectively. In turn, for small angles Eq. (2.37b) can also be written as

$$\frac{\sin\theta_1}{\sin\theta_2} = \frac{k_2'}{k_1'} \tag{2.37c}$$

where k_1' and k_2' can be thought of as modified thermal conductivities. However, the approximate form of Eq. (2.37b) reads as

$$\frac{\sin\theta_1}{\sin\theta_2} \approx \frac{k_2}{k_1} \tag{2.37d}$$

for small angles of incidence and refraction. For constant thermal conductivities, each of the forms contained in Eqs. (2.37a), (2.37b), and (2.37c) can be represented, respectively, as

$$\left(\frac{\partial T_1}{\partial y}\right)_{0^+} + \left(\frac{\partial T_2}{\partial y}\right)_{0^-} = \text{constant}, \tag{2.38a}$$

$$\tan\theta_1 + \tan\theta_2 = \text{constant}, \tag{2.38b}$$

and

$$\sin\theta_1 + \sin\theta_2 = \text{constant}. \tag{2.38c}$$

It is to be noted that the message contained in Eqs. (2.37a), (2.37b), and (2.37c) are but principally one and the same: the very proposition of law of motive force. Further, it is to be noted that Eq. (2.37b) is a consequence of tangent law in heat conduction [66], whereas Eq. (2.37c) is an outcome of Fermat's principle and also modeled through dynamic programming approach [67, 68].

Comparing Eq. (2.37b) with (2.37d) we observe that there is a sacrifice in the degree of accuracy. This criterion of accuracy is to be judged from the pertinent application in question. For example, let us consider the more generalized situation of coupled conductive–convective heat transport mechanism [69]. Approximation of surface heat flux at the solid surface of the form

$$-k_1\left(\frac{\partial T_1}{\partial y}\right)_{0^+} \approx \frac{k_1(\Delta T_1)_{\bar{t}}}{\bar{t}} \tag{2.39}$$

is valid for a linear temperature distribution across the wall according to the following relation:

$$T_1 = T_w(x) + \frac{(\Delta T_1)_{\bar{t}}}{\bar{t}} \tag{2.40}$$

where the subscript w refers to the interfacial condition based upon average thickness $\bar{t}$ of the insulation volume. According to the theory of similarity [70] for a nonlinear temperature variation across the wall we may write

$$\left(\frac{\partial T_1}{\partial y}\right)_{0^+} \approx \varepsilon\frac{(\Delta T_1)_{\bar{t}}}{\bar{t}} \tag{2.41}$$

where ε is a correction factor for the distorted temperature profile. The slope on the right side of Eq. (2.41) is a single-valued function of $\frac{(\Delta T_1)_t}{\bar{t}}$. Equations (2.36a) and (2.41) can be rearranged in the form

$$\frac{(\Delta T_1)_{\bar{t}}}{(\Delta T_2)_{\delta_T}} = \varepsilon\frac{k_2}{k_1}\frac{t}{x}\frac{x}{\delta_T} \tag{2.42}$$

where δ_T is the thermal boundary thickness of the medium. Approximate general local Nusselt number correlation can be expressed in the form [71]

$$Nu_x = \frac{x}{\delta_T} = C\mathrm{Pr}^a\mathrm{Re}_x^b \tag{2.43}$$

where a, b, and C are constants. Thus, the relative temperature drop term $(\Delta T_r)_{\bar{t}}$ contained in Eq. (2.42) is expressible as

$$(\Delta T_r)_{\bar{t}} = \frac{(\Delta T_1)_{\bar{t}}}{(\Delta T_2)_{\delta_T}} = \varepsilon Cf\left(\frac{k_2}{k_1}\frac{t}{x}\mathrm{Pr}^a\mathrm{Re}_x^b\right) \tag{2.44}$$

where Pr is the Prandtl number and Re_x is the local Reynolds number of the flow arrangement. Clearly, $(\Delta T_r)_{\bar{t}}$ is a single-valued function of the parametric group

$$Br_x = \frac{k_2}{k_1}\frac{t}{x}\mathrm{Pr}^a\mathrm{Re}_x^b \tag{2.45}$$

known as local Brun number. This local Brun number criterion [72] determines the degree of accuracy surrendered on solving a conjugate problem as a nonconjugate one. In view of this engineering approximation either of Eqs. (2.37a), (2.37b), (2.37c), or (2.37d) can be quantitatively treated to comply with the law of motive force expressed in its fundamental form as

$$\theta_1 + \theta_2 = \text{constant.} \tag{2.46}$$

However, qualitatively ordinary optical rays obey Riemannian geometry, while thermal rays are described by Finslerian geometry [73]. In Eqs. (2.37b) and (2.37d), it is revealed that between tangent law of heat conduction and Fermat's principle in optics there exists a difference only in the degree of accuracy. Philosophically, they are but one and the same: the unique optimization strategy of nature—the law of motive force. Comparing Eqs. (2.37b) and (2.37c) it can be perceived the tangent law of conductive heat transfer pertaining to a combination of media (k_1, k_2) is equivalent to Fermat's principle of optics to an altered combination of media (k_1', k_2'). Unlike point-to-point flow, the demarcation between Fermat type flow and the constructal law is well established in the relevant literature [63, 74]. The Fermat type principle can be readily recognized as a demonstration of the law of motive force, whereas in constructal law the competition of forward and backward motivation is manifested through slower (diffusion-like) and faster (convection-like) processes. Further, it is also to be observed that the law of motive force, while observed in nature or artificial systems, exhibits a category of equipartition [63, 75–79] principle in some macroscopic domains with finite time and length scale.

2.9 Discussions

A number of mathematical studies on nonstandard methods in the calculus of variations [80] are available. However, the present study is under the proposition of a natural law: the law of motive force. It has been suggested in some authoritative treatises that in many problems where we only want a few values of the nonlinear partial differential equation, we can solve the associated variational problems instead [81]. Application of the law of motive force is a justification of the physical basis in this direction.

Specifically, the law of motive force has been exploited for a class of purely conductive systems, where a limited amount of insulating material is to be distributed over a plane wall or curved surface with arbitrary temperature distributions for minimum heat transfer. The method is also extended to a more generalized situation of a stream suspended in an environment of different temperatures and where the wall temperature distribution is affected by the amount of insulation added. The results obtained are in conformity with those reported in the literature [27, 64]. The equivalence of the result obtained in applying the variational principle for a prescribed temperature history to that obtained for a prescribed heat flux is well established in the relevant literature [82].

From the physics of such class of extremum problems, it has been argued that a truly minimum exists. However, the quantification of minimum heat transfer has not been reported here. Once the optimum profile of insulation is obtained, the minimum heat transfer quantity follows readily from the routine procedure and is available in the literature [27]. Since any distribution pattern of insulting material can be represented by a Fourier series, it has been insinuated that such class of conductive minimum heat transfer problems pertain to a category of the Sturm–Liouville system [83–85].

Finally, from a summation form of the law of motive force formulation, a ratio form is derived when one of the constituent competing mechanisms turns out to be a constant. Thus, the ratio form of law of motive force is more restrictive than its corresponding summation counterpart. It turns out to be a mathematical fact that when the ratio form is valid the summation form is spontaneously granted, but not vice versa. In view of this argument Schmidt's criterion for the fin design, the tangent law of conductive heat transport and the Fermat's law of geometrical optics obeys the law of motive force. The constructal law is realized as a competition between slower (diffusion-like) and faster (convection-like) processes and thus complies with the law of motive force. Hence, the basis for analogies among some physical theories is sought. The fundamental feature of this optimization is but a category of macroscopic organization with a class of equipartition principle [63, 75–79].

References

1. Courant, R., Robbins, H.: What is Mathematics? (Stewart, I., Revised), pp. 329–397. Oxford University Press, Oxford (2007)
2. Hancock, H.: The Theory of Maxima and Minima. Dover, New York (1960)
3. Niven, I., Lance, L.H.: Maxima and Minima Without Calculus. MAA, Washington (1981)
4. Tikhomirov, V.M.: Stories About Maxima and Minima. Mathematical World-I, pp. 3–8. AMS, Rhode Island (1990)
5. Tonti, E.: A systematic approach to the variational formulation in physics and engineering. In: Autumn Course on Variational Methods in Analysis and Mathematical Physics. ICTP, Trieste, 20 Oct–11 Dec 1981
6. Yourgrau, W., Mandelstam, S.: Variational Principles in Dynamics and Quantum Theory, p. 175. Dover, New York (2007)
7. Bejan, A.: Advanced Engineering Thermodynamics, pp. 26–34. Wiley, New York (2006)
8. Müller, I.: A History of Thermodynamics. Springer, New York (2007)
9. Truesdell, C.: The Tragicomedy of Classical Thermodynamics. CISM, Udine, Courses and Lectures, No. 70. Springer, New York (1983)
10. Buchdahl, H.A.: A variational principle in classical thermodynamics. Am. J. Phys. **55**, 81–83 (1987)
11. Hancock, H.: The Theory of Maxima and Minima, pp. 150–151. Dover, New York (1960)
12. Biot, M.A.: Variational Principles in Heat Transfer. Oxford University Press, Oxford (1970)
13. Donnelly, R.J., Herman, R., Prigogine, I. (eds.): Non-Equilibrium Thermodynamics, Variational Techniques and Stability. University of Chicago Press, Chicago (1966)
14. Finlayson, B.A., Scriven, L.E.: On the search for variational principles. Int. J. Heat Mass Transf. **10**, 799–821 (1967)
15. Goldstine, H.H.: A History of the Calculus of Variations from the 17th Through 19th Century. Springer, New York (1980)
16. Sieniutycz, S.: Conservation Laws in Variational Thermo-Hydrodynamics. Springer, New York (1994)
17. Sieniutycz, S.: Progress in variational formulations of macroscopic processes. In: Sieniutycz, S., Farkas, H. (eds.) Variational and Extremum Principles in Macroscopic Systems. Elsevier, London (2004)
18. Todhunter, I.: A History of the Calculus of Variations During the Nineteenth Century. Dover, New York (2005)
19. Yourgrau, W., Mandelstam, S.: Variational Principles in Dynamics and Quantum Theory, pp. 162–180. Dover, New York (2007)
20. Prigogine, I.: Remarks on variational principles. In: Donnelly, R.J., Herman, R., Prigogine, I. (eds.) Non-Equilibrium Thermodynamics, Variational Techniques and Stability. Chicago University Press, Chicago (1966)

21. Kantrovich, L.V., Krylov, V.I.: Approximate Methods of Higher Analysis (trans: Benster, C.D.). Interscience, New York (1964)
22. Schechter, R.S.: Variational principles for continuum systems. In: Donnelly, R.J., Herman, R., Prigogine, I. (eds.) Non-Equilibrium Thermodynamics, Variational Techniques and Stability. Chicago University Press, Chicago (1966)
23. Prigogine, I.: Evolution criteria, variational properties and fluctuations. In: Donnelly, R.J., Herman, R., Prigogine, I. (eds.) Non-Equilibrium Thermodynamics, Variational Techniques and Stability. Chicago University Press, Chicago (1966)
24. Pontryagin, L.S., Boltyanskii, V.G., Gamkrelidze, R.V., Mishchenko, E.F.: The Mathematical Theory of Optimal Processes (trans: Trirogoff, K.N.). In: Neustadt, L.W. (ed.), pp. 1–73, 75–114, 239–256. Wiley-Interscience, New York (1965)
25. Fel'dbaum, A.A.: On the question of synthesizing optimum automatic control systems. In: Transactions of the Second All Union Conference on Automatic Control Theory-II, USSR Academy of Science (1955) (in Russian)
26. Ten Hoor, M.J.: The variational method—why stop half way? Am. J. Phys. **62**, 166–168 (1994)
27. Bejan, A.: How to distribute a finite amount of insulation on a wall with nonuniform temperature. Int. J. Heat Mass Transf. **36**, 49–56 (1993)
28. Bejan, A.: Second-law analysis in heat transfer and thermal design. Adv. Heat Transf. **15**, 1–58 (1982)
29. Bejan, A.: Entropy generation minimization: the new thermodynamics of finite-size devices and finite-time processes. J. Appl. Phys. **79**, 1191–1218 (1996)
30. Bejan, A.: A general variational principle for thermal insulation system design. Int. J. Heat Mass Transf. **22**, 219–228 (1979)
31. Tonti, E.: The reason for analogies between physical theories. Appl. Math. Modell. **1**, 37–50 (1976)
32. Censor, D.: Fermat's principle and real space time rays in absorbing media. J. Phys. A Math. Gen. **10**, 1781–1790 (1977)
33. Newcomb, W.A.: Generalized Fermat principle. Am. J. Phys. **51**, 338–340 (1983)
34. Bejan, A.: Advanced Engineering Thermodynamics, pp. 705–841. Wiley, New York (2006)
35. Bejan, A.: Shape and Structure, from Engineering to Nature. Cambridge University Press, Cambridge (2000)
36. Lemons, D.S.: Perfect Form, pp. ix–xi. Princeton University Press, Princeton (1997)
37. Pramanick, A.K., Das, P.K.: Method of synthetic constraint, Fermat's principle and the constructal law in the fundamental principle of conductive heat transport. Int. J. Heat Mass Transf. **50**, 1823–1832 (2007)
38. Feynman, R.: The Character of Physical Law, pp. 149–173. MIT Press, Massachusetts (1967)
39. Leff, H.S.: What if entropy were dimensionless? Am. J. Phys. **67**, 1114–1122 (1999)
40. Schmidt, E.: Die Wärmeübertragung durch Rippen. Z. Ver. Dt. Ing. **70**, 885–889, 947–951 (1926) (in German)
41. Duffin, R.J.: A variational problem relating to cooling fins. J. Math. Mech. **8**, 47–56 (1959)
42. Jany, P., Bejan, A.: Ernst Schmidt's approach to fin optimization: an extension to fins with variable conductivity and the design of ducts for fluid flow. Int. J. Heat Mass Transf. **31**, 1635–1644 (1988)
43. Clausius, R.: On the motive power of heat, and on the laws which can be deduced from it for the theory of heat (trans: Magie, W.F.). In: Mendoza, E. (ed.) Reflections on the Motive Power of Fire. Dover, New York (2005)
44. Bejan, A.: Heat Transfer, pp. 42–44. Wiley, New York (1993)
45. Feynman, R.: The Character of Physical Law, pp. 84–107. MIT Press, Massachusetts (1967)
46. Rosen, J.: Symmetry in Science, pp. 134–154. Springer, New York (1995)
47. Van Fraassen, B.C.: Laws and Symmetry. Oxford University Press, Oxford (1989)
48. Todhunter, I.: A History of the Calculus of Variations During the Nineteenth Century, pp. 243–253. Dover, New York (2005)

49. Culverwell, E.P.: On the discrimination of maxima and minima solutions in the calculus of variations. Philos. Trans. R. Soc. Lond. A **178**, 95–129 (1887)
50. Fourier, J.: The Analytical Theory of Heat (trans: Freeman, A.), pp. 137–144. Dover, New York (2003)
51. Carslaw, H.S.: Introduction to the Theory of Fourier Series and Integrals, pp. 323–328. Dover, New York (1930)
52. Carslaw, H.S., Jaeger, J.C.: Conduction of Heat in Solids, pp. 6–8. Oxford University Press, Oxford (1959)
53. Tolstov, G.P.: Fourier Series (trans: Silverman, R.A.), pp. 12, 60. Dover, New York (1976)
54. Whittaker, E.T., Watson, G.N.: A Course on Modern Analysis, pp. 224–225. Cambridge University Press, Cambridge (1996)
55. Carslaw, H.S.: Introduction to the Theory of Fourier Series and Integrals, pp. 284–288. Dover, New York (1930)
56. Carslaw, H.S.: Introduction to the Theory of Fourier Series and Integrals, pp. 329–361. Dover, New York (1930)
57. Bridgman, P.W.: Tolman's principle of similitude. Phys. Rev. **8**, 423–431 (1916)
58. Bellman, R.: Methods of Nonlinear Analysis-I, pp. 304–330. Academic Press, New York (1970)
59. Courant, R.: Differential and Integral Calculus-I (trans: McShane, E.J.), pp. 178–179. Wiley, New York (1967)
60. Bellman, R.: Introduction to the Mathematical Theory of Control Processes-I, pp. 33–34. Academic Press, New York (1970)
61. Bejan, A.: Heat Transfer, p. 40. Wiley, New York (1993)
62. Nusselt, W.: Die Abhängigkeit der Wärmeübergangszahl von der Rohrlänge. VDI Z. **54**, 1154–1158 (1910) (in German)
63. Pramanick, A.K., Das, P.K.: Note on constructal theory of organization in nature. Int. J. Heat Mass Transf. **48**, 1974–1981 (2005)
64. Kalyon, M., Sahin, A.Z.: Application of optimal control theory in pipe insulation. Numer. Heat Transf. A- Appl. **41**, 391–402 (2002)
65. Özişik, M.N.: Heat Conduction, pp. 17–20. Wiley, New York (1993)
66. Tan, A., Holland, L.R.: Tangent law of refraction for heat conduction through an interface and underlying variational principle. Am. J. Phys. **58**, 988–991 (1990)
67. Bellman, R.: Dynamic Programming. Dover, New York (2003)
68. Sieniutycz, S.: Dynamic programming approach to a Fermat type principle for heat flow. Int. J. Heat Mass Transf. **43**, 3453–3468 (2000)
69. Pramanick, A.K., Das, P.K.: Heuristics as an alternative to variational calculus for optimization of a class of thermal insulation systems. Int. J. Heat Mass Transf. **48**, 1851–1857 (2005)
70. Sedov, L.I.: Similarity and Dimensional Methods in Mechanics (trans: Kisin, V.I.). Mir, Moscow (1982)
71. Bejan, A.: Convection Heat Transfer, pp. 37–42. Wiley, New York (2004)
72. Luikov, A.V.: Conjugated heat transfer problems. Int. J. Heat Mass Transf. **3**, 293–303 (1961)
73. Janyszek, H., Mrugala, R.: Riemannian and Finslerian geometry and fluctuations of thermodynamic systems. In: Sieniutycz, S., Salamon, P. (eds.) Nonequilibrium Theory and Extremum Principles. Taylor & Francis, New York (1990)
74. Bejan, A.: Constructal comment on a Fermat-type principle for heat flow. Int. J. Heat Mass Transf. **46**, 1885–1886 (2003)
75. Bejan, A.: Advanced Engineering Thermodynamics, pp. 352–356, 464–466, 569–571, 709–721, 782–788, 816–820. Wiley, New York (2006)
76. Bejan, A.: Shape and Structure, from Engineering to Nature, pp. 53–56, 84–88, 99–108, 151–161, 220–223, 234–242, 287–288. Cambridge University Press, Cambridge (2000)
77. Bejan, A., Tondeur, D.: Equipartition, optimal allocation, and the constructal approach to predicting organization in nature. Rev. Gen. Therm. **37**, 165–180 (1998)

78. De Vos, A., Desoete, B.: Equipartition principle in finite-time thermodynamics. J. Non-Equilib. Thermodyn. **25**, 1–13 (2000)
79. Lewins, J.: Bejan's constructal theory of equal potential distribution. Int. J. Heat Mass Transf. **46**, 1541–1543 (2003)
80. Tuckey, C.: Nonstandard Methods in the Calculus of Variations. Pitman Research Notes in Mathematics Series, vol. 297. Longman Scientific & Technical, Essex (1993)
81. Bellman, R.: Selective Computation, p. 38. World Scientific, Philadelphia (1985)
82. Lardner, T.J.: Biot's variational principle in heat conduction. AIAA J. **1**, 196–206 (1963)
83. Courant, R., Hilbert, D.: Methods of Mathematical Physics-I, pp. 291–295. Wiley, Berlin (2008)
84. Hildebrand, F.B.: Methods of Applied Mathematics, pp. 89–92, 145–148. Dover, New York (1992)
85. Morse, P.M., Feshbach, H.: Methods of Theoretical Physics-I, pp. 719–726. McGraw-Hill, New York (1953)

Chapter 3
Conjugate Heat Transport Systems

> *A new scientific truth does not triumph by convincing its opponents and making them see the light, but rather because its opponents finally die, and a new generation grows up that is familiar with it.*
>
> M. Planck

In this chapter, we further employ the law of motive force, a physical principle, to a class of more complicated situation of conductive–convective conjugate heat transfer problems. We provide a complete analytical solution for a classically unsolved problem of generalized Pohlhausen's solution of forced convection with Hartee's velocity profile in relation to the design of thermal insulation systems. Initially, the law of motive force is employed to a nonconjugate heat transfer problem with assumed boundary layer type variation of convective heat transfer coefficient. Next, relaxing this a priori known variation of convective heat transfer coefficient to be unknown, the actual nonlinear profile of insulation thickness for minimum heat transfer from a flat plate is derived by the above natural law alone. The method of intersecting asymptotes is utilized to find an upper ceiling of insulating material beyond which the optimization problem reduces to its triviality. For ease of fabrication, tapered insulation profile over the actual nonlinear one is also considered and analytical solution is provided. Finally, the place of law of motive force among other established methodologies of thermodynamic optimization is discussed.

3.1 The Problem

In the development of a generalized methodology it is crucial that we view apparently antagonistic avenues of facts at least with a good deal of qualitative similarities at the outset. This habit is an eye-opener and makes one to realize every problem on the basis of commonality. In this connection, we will consider the problem of conduction–convection conjugate [1–10] heat transfer problem as an evolved category of more complex conductive heat transfer problems.

In conventional formulation of heat transfer between a stream of fluid and a flat plate, boundary conditions are normally stipulated at the solid–liquid interface, that is, at the top of the plate as shown in Fig. 3.1. Here, x and y are two orthogonal directions, u and v are fluid stream velocities along two orthogonal directions, k_f is

A. K. Pramanick, *The Nature of Motive Force*, Heat and Mass Transfer,
DOI: 10.1007/978-3-642-54471-2_3,

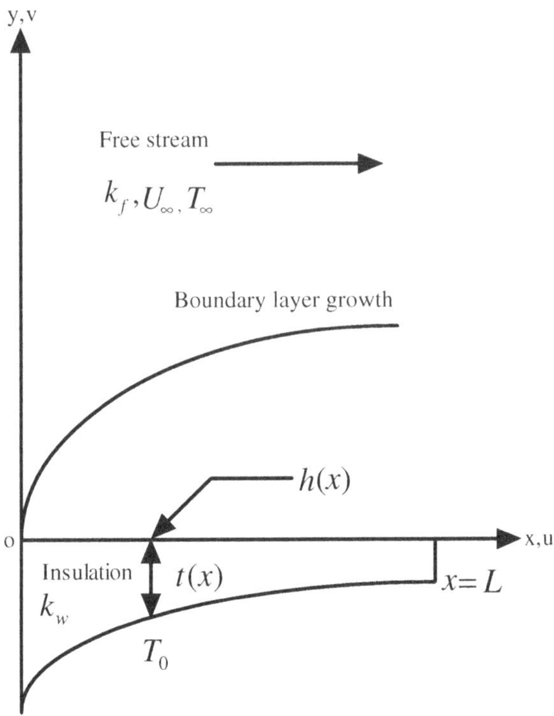

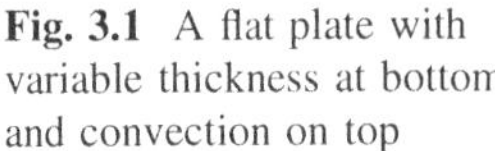
Fig. 3.1 A flat plate with variable thickness at bottom and convection on top

the thermal conductivity of the fluid, T_∞ and U_∞ are free stream temperature and velocity, respectively, at the top of the plate of length L. At the top of the plate the length wise varying convective heat transfer coefficient is $h(x)$. At the bottom of the plate the thermal conductivity of the insulating material is k_w, varying thickness of the insulation volume is $t(x)$ and T_0 is the constant temperature. However, in a large number of applications, the temperature at the bottom surface of the plate is either specified or can be estimated. If the plate is of negligible thickness or has a high thermal conductivity, the temperature drop between the top and bottom surfaces can be neglected and the problem is solved purely in convective heat transfer regime [11]. When thickness of the plate is not negligible or even varying along its length and the thermal conductivity of the wall material is poor, the boundary conditions at the bottom of the plate are to be considered and the whole problem is to be reformulated as a conductive–convective one. This is a fundamental mathematical challenge imposed by the design criterion of thermal insulating systems.

The present discussion is with specific reference to conductive–convective heat transfer along a flat plate of variable thickness [12] and is thus an attempt to generalize Pohlhausen's problem [11]. At present, this classically unsolved problem is solved by applying the law of motive force. In an earlier unsuccessful attempt by Lim et al. [12], the goal of the work was the optimal distribution of a limited quantity of insulating material on the backside of a convectively cooled flat plate.

In the first part of the said paper [12], the authors assumed a boundary layer type variation of convective heat transfer coefficient and determined the total thermal resistance as a series combination contributed by conduction and convection. Accounting for the constraint in insulation volume, the authors [12] could cast the optimization problem in Euler–Lagrange form and obtained an analytical solution. The second part of the aforementioned article [12] solves the same problem of forced convection using the conjugate heat transfer condition without assuming any heat transfer coefficient beforehand. They attempted a general formulation for convective cooling of a flat plate with lagging of arbitrary thickness on the other surface. However, the authors Lim et al. failed to extend the calculus of variations or other methodologies to find out the optimum profile of insulation in this case. Instead of considering the optimum nonlinear profile thickness, they numerically determined the total rate of heat loss for a linearly varying tapered shape and showed it to be less than that of a plate with constant thickness of insulation.

In this chapter, the above problem is revisited. First, it is shown that the result deduced by variational calculus can be obtained by applying the law of motive force grounded on the physics of the problem. Second, the law of motive force is not only applicable to the first part of the problem [12] with known variation of heat transfer coefficient, but can also be employed for the second part where the estimation of heat transfer is based on a truly conjugate formulation. Further, an approximate bound of the insulation volume is provided by Bejan's method of intersecting asymptotes for any meaningful optimization of the insulation design. An analytical treatment is extended for the design of tapered insulation profile. Finally, it is argued that the said problem demonstrates a category of equipartition. The chapter explains the result obtained by the author [13] in view of the proposed law of motive force.

3.2 The Physical Model

A flat plate with variable thickness and finite length is considered as shown in Fig. 3.1. The bottom of the plate is exposed to an environment with high convective heat transfer coefficient such that the temperature of the surface remains practically uniform at T_0. The top and flat sides of the plate are in thermal communication with a different flow characteristic $\left(k_f, U_\infty, T_\infty\right)$. The total inventory of the wall material is fixed. It is desirable to seek an optimal distribution of the wall material to achieve minimum heat transfer from the plate [12]. This is a fundamental optimization problem encountered in the design of thermal insulating systems.

In this chapter, we assume that the driving potential for heat transfer remains piecewise constant, i.e.,

$$T_\infty - T_0 = \Delta T = \text{constant}. \tag{3.1}$$

This concept of uniformed (equipartitioned) potential difference is due to [14]. The material volume per unit length is a constant and can be expressed in terms of average thickness as

$$\bar{t} = \frac{1}{L}\int_{0}^{L} t(x)\mathrm{d}x \tag{3.2a}$$

or

$$\int_{0}^{1} \frac{t(\xi)}{\bar{t}}\mathrm{d}\xi = 1, \text{ with } \xi = \frac{x}{L} \tag{3.2b}$$

where $\bar{t}$ is the length-based average thickness of insulation distribution and ξ is the nondimensional length of the flat plate.

3.3 Optimization with Assumed Variation of Heat Transfer Coefficient

In a forced laminar convection [15–21] heat transfer from a flat plate, it is legitimate to assume a power law variation of heat transfer coefficient along the direction of the flow in the form [12]

$$h = h_L \xi^{-n} \tag{3.3}$$

where h_L is the lowest value of convective heat transfer coefficient at the extreme downstream $x = L$ and n is an exponent. Local heat flux q'' driven by temperature potential ΔT can be expressed considering convective and conductive resistances in series as

$$q'' = \frac{\Delta T}{\frac{t(x)}{k_w} + \frac{1}{h(x)}}. \tag{3.4}$$

Equation (3.4) can be integrated for the entire length of the plate invoking Eq. (3.3) to obtain total heat transfer rate q' per unit length perpendicular to the plane in the dimensionless form as

$$\frac{q'}{k_w L \Delta T/\bar{t}} = \int_{0}^{1} \frac{d\xi}{\frac{t}{\bar{t}} + \frac{\xi^n}{\mathrm{Bi}}} \tag{3.5}$$

where Biot number Bi is expressed as

$$\mathrm{Bi} = \frac{h_L \bar{t}}{k_w}. \tag{3.6}$$

Equation (3.6) represents a competition between convection and conduction. Lim et al. [12] constructed an aggregate integral combining Eqs. (3.5) and (3.2b) through a Lagrange multiplier [22]. Finally, the Euler–Lagrange equation [23] of the integral was solved to find the optimum thickness of insulation.

In this monograph, the problem is approached through the law of motive force. We now paraphrase the law of motive force on a contextual basis. As the goal is to reduce the heat loss from the total length of the plate, it is instructive to provide the maximum thickness of insulation where the coefficient of convective heat transfer is maximum. In other words, one should equip the highest conductive resistance where the convective resistance is minimum. This exercise should be carried out for the entire plate length under the constraint of limited insulation material. Logically, this exertion can terminate only when uniform (equipartitioned) total thermal resistance (conductive plus convective) prevails throughout the length of the plate. Clearly, we identify two competing mechanisms as convection and conduction which are regarded, respectively, as forward and backward motivations of the system with respect to the motive of heat transfer. Mathematically, this translates into the equation with the stipulation that the denominator of the integrand in Eq. (3.5) is constant, i.e.,

$$\frac{t}{\bar{t}} + \frac{\xi^n}{\mathrm{Bi}} = \bar{R} = \text{constant}. \tag{3.7}$$

The demonstration of law of motive force is schematically supplemented in Fig. 3.2.

Providing an expression for $t(\xi)/\bar{t}$ in Eq. (3.2b) from Eq. (3.7), we obtain another expression for total resistance as

$$\frac{1}{(n+1)\mathrm{Bi}} + 1 = \bar{R}. \tag{3.8}$$

Eliminating $\bar{R}$ between Eqs. (3.7) and (3.8) leads to the functional form of optimal distribution of insulation thickness t_* as

$$\frac{t_*}{\bar{t}} = 1 + \frac{1}{\mathrm{Bi}}\left(\frac{1}{n+1} - \xi^n\right) \tag{3.9}$$

where Biot number Bi is defined in Eq. (3.6).

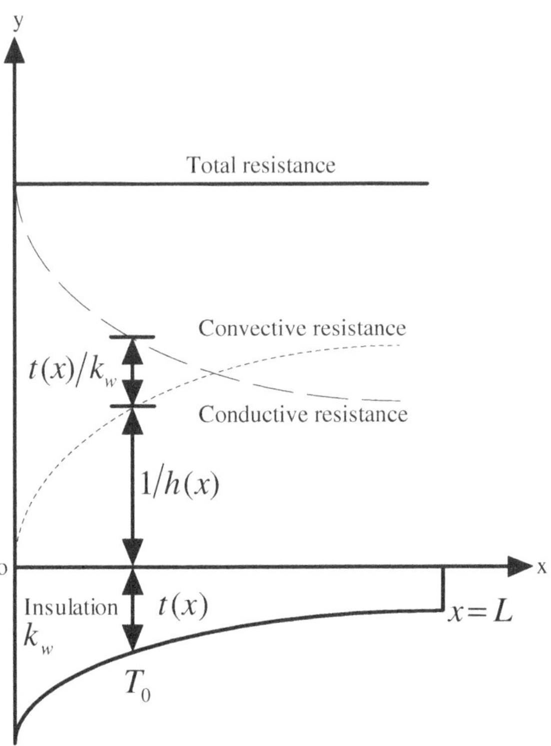

Fig. 3.2 The conceptual basis of law of motive force

Minimum heat transfer rate $q'_{*\min}$ from Eq. (3.5), invoking Eq. (3.9) in the denominator of the integrand, is obtained in a nondimensional form as

$$\frac{q'_{*\min}}{k_w L \Delta T/\bar{t}} = \frac{\text{Bi}}{\text{Bi} + (n+1)^{-1}}. \tag{3.10}$$

Equations (3.9) and (3.10) are the important results for the optimum allocation of insulation and were obtained employing calculus of variations in the open literature [12].

3.4 Optimization with Unknown Variation of Convective Heat Transfer Coefficient

In actual practice, neither the variation of temperature at the top surface of the plate nor convective heat transfer coefficient is known a priori [12], rather it is to be determined from one of conjugate convective–conductive formulation of the problem.

At the solid–fluid interface, the formulations ordinarily belong to the third kind of boundary conditions involving heat transfer coefficient calculated beforehand. It is worth recognizing that the third kind of boundary conditions are not valid for many cases since they lead to contradictory or even physically unrealistic results [24, 25].

Neglecting dissipation, for Eckert number Ec much less than unity ($Ec \ll 1$), boundary layer energy equation can be written in the form [26]

$$\frac{d^2\theta}{d\eta^2} + \frac{1}{2}\Pr f \frac{d\theta}{d\eta} = 0 \tag{3.11}$$

with the definitions

$$\eta = \frac{y}{\sqrt{vx/U_\infty}} = \frac{y}{x}\mathrm{Re}_x^{1/2}, \theta(\eta) = \frac{T_f - T_0}{T_\infty - T_0}, \text{ and } \frac{df}{d\eta} = \frac{u}{U_\infty} \tag{3.12}$$

where T_f is the temperature of the fluid, v is the kinematic viscosity, Pr is the Prandtl number, and Re_x is the local Reynolds number of the fluid flow conditions. The similarity function $f(\eta)$ is obtained from the momentum equation of Blasius form [27–35]. Free stream boundary condition by definition reads as

$$\theta \to 1 \text{ at } \eta \to \infty \; (y \to \infty). \tag{3.13}$$

Considering that maximum thickness of the wall is much smaller than its length, longitudinal conduction through the insulating material can be neglected. Thus, the conjugate boundary condition can be modeled at the interface $y = 0$ using the fourth kind of boundary condition [36] as

$$k_f\left(\frac{\partial T_f}{\partial y}\right)_0 = k_w\left(\frac{T_w - T_0}{t(x)}\right)_0 \tag{3.14a}$$

and

$$T_f = T_w \tag{3.14b}$$

where T_w is the temperature of insulating wall at the solid–liquid interface.

The nondimensional version of these two boundary conditions (3.14a) and (3.14b) become

$$J\frac{\partial\theta}{\partial\eta} = \theta \text{ at } \eta = 0 \tag{3.15}$$

where

$$\bar{J} = \frac{k_f}{k_w}\frac{\bar{t}}{L}\mathrm{Re}_L^{1/2} \tag{3.16}$$

and

$$J = \bar{J}\frac{t}{\bar{t}}\xi^{-1/2}. \tag{3.17}$$

The dimensionless number J is in general a function of x except for some special functional form of $t(x)$ and represents dimensionless insulation volume. The quantity $\bar{J} \to 0$ represents Pohlhausen's limit, that is, for the isothermal plate with negligible wall thickness.

Equation (3.11) can be integrated in a straightforward manner using the relation (3.17) and boundary conditions (3.13) and (3.15) to yield

$$\theta'(0) = \left\{ \bar{J}\frac{t}{\bar{t}}\xi^{-1/2} + \int_0^\infty \exp\left[-\frac{\Pr}{2}\int_0^\beta f(\alpha)\mathrm{d}\alpha \right] \mathrm{d}\beta \right\}^{-1} \tag{3.18}$$

where α and β are two dummy variables. But the improper integral in the denominator of Eq. (3.18) is well known in the literature [11] and for $\Pr > 0.5$ it is most accurately correlated as

$$\int_0^\infty \exp\left[-\frac{\Pr}{2}\int_0^\beta f(\alpha)\mathrm{d}\alpha \right] \mathrm{d}\beta = \left(0.332\ \Pr^{1/3}\right)^{-1}. \tag{3.19}$$

Our concern is to calculate the overall heat transfer rate through the entire length of the plate using the relation

$$q' = \int_0^L k_f \left(\frac{\partial T}{\partial y}\right)_{y=0^+} \mathrm{d}x = \Delta T k_f \mathrm{Re}_L^{1/2} \int_0^1 \theta'(0)\xi^{-1/2}\mathrm{d}\xi, \tag{3.20}$$

etc., where Re_L is the Reynolds number at the extreme downstream of the flat plate. Invoking Eqs. (3.18) and (3.19) into Eq. (3.20) a nondimensional equation for heat transfer is resulted as

$$\frac{q'}{k_w L \Delta T/\bar{t}} = \int_0^1 \frac{d\xi}{\frac{t}{\bar{t}} + \frac{\xi^n}{\bar{J}h_L}} \tag{3.21}$$

$$\text{with } n = 1/2 \text{ and } h_L = 0.332\ \Pr^{1/3} \text{ in particular.} \tag{3.22}$$

It may be noted that Eq. (3.21) has a form exactly equivalent to that of Eq. (3.5). Therefore, the same law of motive force can be extended and the optimum variation of insulation thickness can be determined readily.

It may further be observed that even for this purely conjugate heat transfer situation, one can exploit the calculus of variations to obtain the optimum profile for insulation thickness. Using Eqs. (3.21) and (3.2b), one may formulate a problem of unconstrained optimization with the introduction of a Lagrange multiplier [22] as

$$\Phi = \int_0^1 \left(\frac{1}{\frac{t(\xi)}{\bar{t}} + \frac{\xi^n}{\bar{J}h_L}} + \lambda \frac{t(\xi)}{\bar{t}} \right) \mathrm{d}\xi = \int_0^1 F \mathrm{d}\xi \tag{3.23}$$

where the factor λ is a Lagrange multiplier, F is the shorthand for the integrand, and Φ is the aggregate integral. The optimal thickness is the solution of the following Euler–Lagrange [23] equation:

$$\frac{\partial F}{\partial t} - \frac{\mathrm{d}}{\mathrm{d}\xi}\left[\frac{\partial F}{\partial(\mathrm{d}t/\mathrm{d}\xi)}\right] = 0. \tag{3.24}$$

Since, the integrand in Eq. (3.23) is independent of the slope of the profile, Eq. (3.24) takes a simple look at

$$\frac{\partial F}{\partial t} = 0. \tag{3.25}$$

The resulting expression for optimal thickness distribution involving Lagrange multiplier stands as

$$\frac{t(\xi)}{\bar{t}} = (\lambda)^{-1/2} - \frac{\xi^n}{\bar{J}h_L}. \tag{3.26}$$

From the volume constraint (3.2b), we obtain another expression for the parameter $(\lambda)^{-1/2}$ as

$$(\lambda)^{-1/2} = 1 + \frac{1}{(n+1)\bar{J}h_L}. \tag{3.27}$$

Combining Eqs. (3.26) and (3.27), we conclude with the expression for optimum insulation profile t_{opt} as

$$\frac{t_{\mathrm{opt}}}{\bar{t}} = 1 + \frac{1}{\bar{J}h_L}\left(\frac{1}{n+1} - \xi^n\right). \tag{3.28}$$

Employing this profile shape, nondimensionalized minimum heat transfer $q'_{\min}$ from Eq. (3.21) reads as

$$\frac{q'_{\min}}{k_w L\Delta T/\bar{t}} = \frac{\bar{J}h_L}{\bar{J}h_L + (n+1)^{-1}}. \tag{3.29}$$

One may check that the application of the law of motive force also provides the same result as given in Eqs. (3.28) and (3.29).

3.5 Bounds of Insulation Volume

It is implied that optimization for minimum heat transfer is a worthy endeavor only when an amount of insulating material falls within a limit. To bracket this limit, one can integrate Eq. (3.21) for two different extreme conditions.

When there is an acute scarcity of insulating material passing to the lower limit $\bar{J} \to 0$ we have

$$\underset{\bar{J}\to 0}{\mathrm{Lt}}\left(\frac{q'}{k_w L\Delta T/\bar{t}}\right) = \underset{\bar{J}\to 0}{\mathrm{Lt}}\left[\bar{J}\int_0^1 \frac{\mathrm{d}\xi}{\bar{J}\frac{t}{\bar{t}} + \frac{\xi^{1/2}}{0.332\,\mathrm{Pr}^{1/3}}}\right] = 0.664\,\mathrm{Pr}^{1/3}\bar{J}. \tag{3.30}$$

This is the classical Pohlhausen solution [11] with no thickness of the wall or having high conductivity of the wall material.

On the other hand, for overabundance of insulating material, optimization for the profile shape is trivial, that is, $t \to \bar{t}$. Passing to the higher limit $\bar{J} \to \infty$ we obtain

$$\underset{\bar{J}\to \infty}{\mathrm{Lt}}\left(\frac{q'}{k_w L\Delta T/\bar{t}}\right) = \underset{\bar{J}\to \infty}{\mathrm{Lt}}\left[\bar{J}\int_0^1 \frac{\mathrm{d}\xi}{\bar{J}\frac{t}{\bar{t}} + \frac{\xi^{1/2}}{0.332\,\mathrm{Pr}^{1/3}}}\right] = 1. \tag{3.31}$$

Now, we are positioned to fix an upper ceiling for the insulating material using Bejan's method of intersecting asymptotes [37–43]. Elimination of q' term between Eqs. (3.30) and (3.31) yields upper ceiling of insulation volume $\bar{J}_{\max}$ as

$$\bar{J}_{\max} = 1.506\,\mathrm{Pr}^{1/3}. \tag{3.32}$$

In Eq. (3.32) it is revealed that $\bar{J}_{\max}$ scales with $\mathrm{Pr}^{-1/3}$ and bounded in the domain $0 < \bar{J}_{\max} \leq 1.506\,\mathrm{Pr}^{-1/3}$, when optimization problem actually becomes a nontrivial one.

From definition (3.16), it is evident that the parameter $\bar{J}$ represents a competition between convection through the boundary layer and conduction through the insulating material. The value of the parameter $\bar{J}$ of the order of unity signifies a transition between an overall resistance dominated by the insulating material and that of boundary layer. Thus, it is more realistic to treat the limit $\bar{J} \to 0$ as $\bar{J} \ll 1$ and $\bar{J} \to \infty$ as $\bar{J} \gg 1$. From Eqs. (3.22) and (3.28) for nonzero wall thickness it can be read that $\bar{J} \sim 1$. This final result is in quantitative agreement with that obtained in the document [12] after elaborate numerical computations.

3.6 Insulation with Tapered Profile

It has been mentioned earlier that Lim et al. [12] assumed a tapered profile of the insulation and numerically solved the convective heat transfer problem with conjugate boundary condition at the top surface of the plate. Lim et al. selected an insulation profile qualitatively similar to an optimum one, as they could not extend the calculus of variations or else for the insulation problem with a rigorous conjugate boundary condition. With the background provided in Sect. 3.4 such an assumption is not mandatory for the optimum design of insulation. However, a tapered profile of insulation is still of interest due to the ease of fabrication.

We show here that the analysis presented in Sect. 3.4 is in general enough to handle the taper profile of the insulation and a closed form of expression can be deduced for the minimum heat transport rate. It can be noticed that for $n = 1$ Eq. (3.28) represents a tapered profile for the distribution of insulating material. The reciprocal of the group $\bar{J}h_L$ is termed as taper parameter. With these, Eq. (3.28) resumes a linearized form for the tapered profile t_{taper} as

$$\frac{t_{\text{taper}}}{\bar{t}} = 1 + b\left(\frac{1}{2} - \xi\right), \text{ where } 0 < b = \frac{1}{\bar{J}h_L} \leq 2. \tag{3.33}$$

The expression for heat transfer rate q'_{taper} with this profile is readily obtained from Eq. (3.21) in dimensionless form as

$$\frac{q'_{\text{taper}}}{k_w L \Delta T/\bar{t}} = 2\bar{J}\int_0^1 \frac{\xi \mathrm{d}\xi}{A\xi^2 + B\xi + C} \tag{3.34}$$

$$\text{where } A = -\bar{J}b,\ B = \left(0.332\ \mathrm{Pr}^{1/3}\right)^{-1}, \text{ and } C = \bar{J}\left(1 + \frac{b}{2}\right). \tag{3.35}$$

For all possible practical set of values of the parameters

$$4AC - B^2 = -4\bar{J}^2 b\left(1 + \frac{b}{2}\right) - \left(0.332\ \mathrm{Pr}^{1/3}\right)^{-2} < 0. \tag{3.36}$$

Thus, the algebraic expression for heat transfer rate assumes the form

$$\frac{q'_{\mathrm{taper}}}{k_w L\Delta T/\bar{t}} = 2\bar{J}\ln\left[\left(\frac{A+B+C}{C}\right)^{1/2A}\left(\frac{2A+B+\sqrt{B^2-4AC}}{2A+B-\sqrt{B^2-4AC}}\cdot\frac{B-\sqrt{B^2-4AC}}{B+\sqrt{B^2-4AC}}\right)^{B/2A\sqrt{B^2-4AC}}\right]. \tag{3.37}$$

This is the exact solution of the numerical result presented in [12]. Comparing heat transfer results for the representative material volume $\bar{J} = 1$, thermophysical property $\mathrm{Pr} = 1$, and the optimum taper parameter $b = 2$, one can verify the relative figure of merit from the ratio

$$\frac{q'_{\mathrm{taper}}}{q'_{\mathrm{min}}} = 1.0095. \tag{3.38}$$

The last relation reveals that only 0.1 % improvement is experienced by the actual optimum profile in place of approximated linearized profile. This taper profile is one such among many other competing designs.

In case of constant wall thickness the taper parameter reduces to zero. The heat transfer q'_{constant} can be obtained in algebraic form by evaluating the following reduced integral:

$$\frac{q'_{\mathrm{constant}}}{k_w L\Delta T/\bar{t}} = 2\bar{J}\int_0^1 \frac{\xi \mathrm{d}\xi}{B\xi + \bar{J}}. \tag{3.39}$$

The final algebraic expression takes the form

$$\frac{q'_{\mathrm{constant}}}{k_w L\Delta T/\bar{t}} = 2\bar{J}\ln\left[e^{1/B}\left(\frac{\bar{J}}{B+\bar{J}}\right)^{\bar{J}/B^2}\right]. \tag{3.40}$$

Comparative goodness of tapered profile over uniform thickness can be judged by combining the expressions for heat transfer contained in Eqs. (3.37) and (3.40) as

$$\frac{q'_{\mathrm{taper}}}{q'_{\mathrm{constant}}} = \ln\left[\left(\frac{A+B+C}{C}\right)^{1/2A}\left(\frac{2A+B+\sqrt{B^2-4AC}}{2A+B-\sqrt{B^2-4AC}}\cdot\frac{B-\sqrt{B^2-4AC}}{B+\sqrt{B^2-4AC}}\right)^{B/2A\sqrt{B^2-4AC}}\right] \Bigg/ \ln\left[e^{1/B}\left(\frac{\bar{J}}{B+\bar{J}}\right)^{\bar{J}/B^2}\right]. \tag{3.41}$$

It is easy to verify that this ratio is always less than unity for any value of the design parameter b in the bound $[0, 2]$.

3.7 Law of Motive Force and Commonality of Nature of Optimizations

We will now take a second look at Eq. (3.7) obtained by applying the law of motive force. Substituting Eq. (3.9) into Eq. (3.7) produces an estimate for total resistance, which is exactly the same as that of Eq. (3.8). Again, Eq. (3.9) is the general result of variational principle of optimization [12]. Equation (3.8) was obtained directly from the material volume constraint (3.2a) and the law of motive force (3.7). This proves the worth of postulating the auxiliary constraint (3.7). The synthesis of this supplementary equation contains the whole physics of the problem. Here, the law of motive force presupposes that total conductive and convective resistance is "conserved" though they may not take an equal share at each and every point of the geometry under consideration.

Another pertinent example of the law of motive force is the Bernoulli equation for a stream tube in an inviscid flow field where pressure, kinetic, and potential energies of the flow compete with each other with the stipulation that kinetic energy is the forward motivation; pressure and potential energy belong to backward motivation. The isopotential line provides a basis for understanding the laminar to turbulent transition mechanism as a parallelism between viscid to inviscid transformation [44]. In rigid body mechanics, dropping the pressure term, we obtain the conservation equation for kinetic and potential energy. At some point of the trajectory, the contributing competing components of a constraint may take an equal share. However, this is not a necessary condition for optimality (minimum, shortest, quickest, etc.). Existence of isoline is the only rudimental feature of extremality.

In the heat transfer literature [45, 46], there is perhaps more misunderstanding than real conflict between power maximization (PM) and entropy generation minimization (EGM) line of optimization. All results obtained otherwise can be reproduced by minimizing the entropy production rate. The concept of isoline can still be invoked in the following manner. Minimizing entropy generation rate $\dot{S}_{gen}$ with respect to some design variable χ, the first-order condition for extrema stands as

$$\frac{d}{d\chi}\left(\dot{S}_{gen}\right) = 0. \tag{3.42}$$

Thus, for local thermodynamic equilibrium [47] model in some domain of χ we actually have a pseudo constraint

$$\dot{S}_{\text{gen}} = \bar{\dot{S}}_{\text{gen}} = \text{constant}. \tag{3.43}$$

But Eq. (3.43) constitutes the locus of an isoline and can be deployed with other physical constraints of the model to obtain the condition for optimum. In Eq. (3.43), the competition among forward motivation and backward motivation adds up to a constant on following the law of motive force. This logical foundation constructs the geometrical interpretation of the optimized results. For an ideally reversible process this constant is identically zero.

After identification of m different competing mechanisms $\chi_1, \chi_2, \ldots,$ and χ_m as a class of forward as well as backward motivation, the law of motive force can be laid down as

$$\sum_{i=1}^{m} \chi_i = \bar{\chi} \tag{3.44}$$

where the constant $\bar{\chi}$ is dictated by the finite-time and finite-resources accessible for a system. For a single contributing mechanism, entropy generation between parts of the system can be considered to discover the forward and backward motivation. It has been deduced [48] from purely theoretical reasoning that distribution of driving forces that minimize the entropy is uniform throughout the system for a single acting irreversibility factor. In the literature, such monotonous distribution of physical or nonphysical entities are recognized as principle of equipartition [14, 48–51].

3.8 Discussions

The law of motive force for the design problem of thermal insulating systems is explored. The principle can be considered as general and in specific, methodological replacement for the formal method of calculus of variations. It is founded on easily perceptible logic and employs a few simple mathematical steps to arrive at the final result.

Closed-form expressions for optimum distribution of insulating material for minimum heat transfer from a flat plate when the other side is in convective thermal communication with a forced laminar stream have been obtained. Optimum shape of the profile constitutes an isoline where total resistance contributed by conduction and convection remains uniform throughout the length of the plate. Optimized results are in conformity with the principle of equipartition. Heat transfer results are normalized by the quantity $k_w L \Delta T / \bar{t}$, in which thickness of wall dominates total resistance and provides an effective insulation.

An analytical expression is also derived for tapered insulation profile. However, for certain ranges of the parameter, the optimum solution exhibits only a marginal improvement over the taper profile. Finally, it goes without saying that any optimization problem plays a meaningful role only when resource is limited. An upper ceiling for the insulating material is prescribed beyond which optimization problem is of no challenge.

References

1. Aleksashenko, V.A.: Analytical solution of some conjugated problems of convective heat exchange. Thesis, Minsk (1969)
2. Dorfman, A.S.: Conjugate Problems in Convective Heat Transfer. CRC, Boca Raton (2010)
3. Luikov, A.V., Aleksashenko, V.A., Aleksashenko, A.A.: Analytical methods of solution of conjugated problems in heat transfer. Int. J. Heat Mass Transf. **14**, 1047–1056 (1971)
4. Luikov, A.V.: Conjugate convective heat transfer problems. Int. J. Heat Mass Transf. **17**, 257–265 (1974)
5. Luikov, A.V.: Heat and Mass Transfer (trans: Kortneva, T.), pp. 334–378. Mir, Moscow (1980)
6. Nakayama, A., Koyama, H.: An approximate solution procedure for laminar free and forced convection heat transfer problems. Int. J. Heat Mass Transf. **26**, 1721–1726 (1983)
7. Payvar, P.: Convective heat transfer to laminar flow over a flat plate of finite thickness. Int. J. Heat Mass Transf. **20**, 431–433 (1977)
8. Perelman, T.L.: On conjugated problems of heat transfer. Int. J. Heat Mass Transf. **3**, 293–303 (1961)
9. Pop, I., Ingham, D.B.: A note on conjugate forced convection boundary-layer flow past a flat plate. Int. J. Heat Mass Transf. **15**, 3873–3876 (1993)
10. Pozzi, A., Lupo, M.: The coupling of conduction with forced convection over a flat plate. Int. J. Heat Mass Transf. **23**, 1207–1214 (1989)
11. Pohlhausen, E.: Der Wämeaustausch zwischen festen Körpern und Flüssigkeiten mit kleiner Reibung und kleiner Wärmeleitung. Z. Angew. Math. Mech. **1**, 115–121 (1921) (in German)
12. Lim, J.S., Bejan, A., Kim, J.H.: The optimal thickness of a wall with convection on one side. Int. J. Heat Mass Transf. **35**, 1673–1679 (1992)
13. Pramanick, A.K., Das, P.K.: Heuristics as an alternative to variational calculus for optimization of a class of thermal insulation systems. Int. J. Heat Mass Transf. **48**, 1851–1857 (2005)
14. Pramanick, A.K., Das, P.K.: Note on constructal theory of organization in nature. Int. J. Heat Mass Transf. **48**, 1974–1981 (2005)
15. Eckert, E.R.G.: Die Berechnung des Wärmeübergänges in der Laminaren Grenzschicht um strömter Körper. VDI Forsch. **416**, 1–24 (1942) (in German)
16. Falkner, V.M., Skan, S.W.: Some approximate solutions of the boundary layer equations. Philos. Mag. **12**, 865–896 (1931)
17. Hartee, D.R.: On an equation occurring in Falkner and Skan's approximate treatment of the equations of the boundary layer. Proc. Camb. Phil. Soc. **33**, 233–239 (1937)
18. Lighthill, M.J.: Contributions to the theory of heat transfer through a laminar boundary layer. Proc. R. Soc. Lond. A **202**, 359–377 (1950)
19. Merk, J.H., Prins, J.A.: Thermal convection in laminar boundary layers-I, II, III. Appl. Sci. Res. **4**, 11–24, 195–206, 207–221 (1954)
20. Pletcher, R.H.: External flow forced convection. In: Kakac, S., Shah, R.K., Aung, W. (eds.) Handbook of Single Phase Convective Heat Transfer. Wiley, New York (1987)
21. Weyl, H.: On the differential equations of the simplest boundary layer problems. Ann. Math. **43**, 381–407 (1942)
22. Lewins, J.D.: Introducing the Lagrange multiplier to engineering mathematics. Int. J. Mech. Eng. Ed. **22**, 191–207 (1994)

23. Moiseiwitsch, B.L.: Variational Principles, pp. 14–16. Dover, New York (2004)
24. Chapman, D.R., Rubesin, M.W.: Temperature and velocity profiles in the compressible laminar boundary layer with arbitrary distribution of surface temperature. J. Aeronaut. Sci. **16**, 547–565 (1949)
25. Schlichting, H.: Der Wärmeübergang an einer linearen längsangeströmten ebene: Platte mit veränderlicher Wandttemperatur. Forsch. Geb. Ing. Wes. **17**, 1–7 (1951) (in German)
26. Eckert, E.R.G., Drake Jr., R.M.: Analysis of Heat and Mass Transfer, p. 312. McGraw-Hill, New York (1972)
27. Bejan, A.: Convection Heat Transfer, pp. 49–51, 83. Wiley, New York (2004)
28. Blasius, H.: Grenzschichten in Flüssigkeiten mit kleiner Reibung. Z. Math. Phys. **56**, 1–37 (1908) (in German). Also in NACA TM-1256 (1950)
29. Dyke, M.V.: Perturbation Methods in Fluid Mechanics, pp. 129–132. Parabolic Press, Stanford (1975)
30. Goldstein, S. (ed.): Modern Developments in Fluid Dynamics-I, pp. 135–139. Dover, New York (1965)
31. Görtler, H.: A new series for the calculation of steady laminar boundary layer flows. J. Math. Mech. **6**, 1–66 (1957)
32. Hiemenz, K.: Die Grenzschicht an einem in den gleichförmigen Flüssigkeitsstrom eingetanchten geraden Kreiszylinder. Dissertation, Göttingen University (1911) (in German)
33. Howarth, L.: On the solution of the laminar boundary layer equations. Proc. R. Soc. Lond. A **164**, 547–579 (1938)
34. Rosenhead, L. (ed.): Laminar Boundary Layers, pp. 223–226. Oxford University Press, Oxford (1963)
35. Schlichting, H., Gersten, K.: Boundary Layer Theory, pp. 184–186. Springer, New York (2000)
36. Isachenko, V.P., Osipova, V.A., Sukomel, A.S.: Heat Transfer (trans: Semyonov, S.), pp. 33–34. Mir, Moscow (1987)
37. Bejan, A.: Convection Heat Transfer, pp. 136–141, 211–214, 225–228, 279–286, 404–406, 591–593, 613–616. Wiley, New York (2004)
38. Bejan, A.: Shape and Structure, from Engineering to Nature, pp. 29–41, 45–49, 163–174. Cambridge University Press, Cambridge (2000)
39. Bejan, A., Dincer, I., Lorente, S., Miguel, A.F., Reis, A.H.: Porous and Complex Flow Structures in Modern Technologies, pp. 58–66, 201–212. Springer, New York (2004)
40. Bejan, A., Lorente, S.: Design with Constructal Theory, pp. 81–96, 364–369. Wiley, New York (2008)
41. Lewins, J.: Bejan's constructal theory of equal potential distribution. Int. J. Heat Mass Transf. **46**, 1541–1543 (2003)
42. Nield, D.A., Bejan, A.: Convection in Porous Media, pp. 275–282. Springer, New York (2006)
43. Sadeghipour, M.S., Razi, Y.P.: Natural convection from a confined horizontal cylinder: the optimal distance between the confining walls. Int. J. Heat Mass Transf. **44**, 367–374 (2001)
44. Bejan, A.: Entropy Generation Through Heat and Fluid Flow, pp. 61–62. Wiley, New York (1982)
45. Bejan, A.: Advanced Engineering Thermodynamics, pp. 101–144. Wiley, New York (2006)
46. Bejan, A.: Models of power plants that generate minimum entropy while operating at maximum power. Am. J. Phys. **64**, 1054–1059 (1996)
47. Bejan, A.: Advanced Engineering Thermodynamics, pp. 63–64. Wiley, New York (2007)
48. Bejan, A., Tondeur, D.: Equipartition, optimal allocation, and the constructal approach to predicting organization in nature. Rev. Gen. Therm. **37**, 165–180 (1998)
49. Bejan, A.: Advanced Engineering Thermodynamics, pp. 352–356, 464–466, 569–571, 709–721, 782–788, 816–820. Wiley, New York (2006)
50. Bejan, A.: Shape and Structure, from Engineering to Nature, pp. 53–56, 84–88, 99–108, 151–161, 220–223, 234–242, 287–288. Cambridge University Press, Cambridge (2000)
51. De Vos, A., Desoete, B.: Equipartition principle in finite-time thermodynamics. J. Non-Equilib. Thermodyn. **25**, 1–13 (2000)

Chapter 4
Fluid Flow Systems

So how do you go about teaching them something new? By mixing what they know with what they don't know. Then, when they see in their fog something they recognize they think, "Ah I know that!" And then it's just one more step to "Ah, I know the whole thing." And their mind thrusts forward into the unknown and they begin to recognize what they don't know before and they increase their powers of understanding.

P. Picasso

In this chapter, we focus on the physics of the flow systems with reference to fluid elements. The law of motive force for the flow physics is explored in point-to-point and volume-to-volume flow situations. It attempts to enunciate a clear demarcation between the constructal theory, Fermat's principle, and the law of motive force. The effect of gravity is considered in the formulation of flow geometry employing the law of motive force. Also, the phenomenon of hydraulic jump is theoretically predicted for the first time from the law of motive force. Supporting the buckling theory of fluid jet we predict the fundamental geometric shape of the shear flow and finally strike a unification between the rectangular and triangular shape as the elemental building blocks of self-organized and engineered systems. In sum, we recognize as one of the characteristics of constructal law as well as the law of motive force, the principle of macroscopic equipartition as an authentic basis of design methodology.

4.1 The Problem

The method of thermodynamic optimization has been persuaded under several directions, such as entropy generation minimization [1, 2], exergy analysis [3, 4], finite-time thermodynamics [5, 6], power maximization [7, 8], thermoeconomics [9, 10], and control thermodynamics [11–14]. Recent advancement in thermodynamic optimization is presented here with reference to the generation of optimal geometric forms (topology) in flow systems. The flow configuration has the flexibility to alter its shape and structure. The motive force that governs the generation of geometric form is the result of a conflict between the forward and backward motivation in the pursuit of minimum flow resistance criterion, which in turn is a constant depending on the constitution and make-up of the system. The imposed constraint is global finiteness: volumetric flow rate, weight of the fluid,

A. K. Pramanick, *The Nature of Motive Force*, Heat and Mass Transfer,
DOI: 10.1007/978-3-642-54471-2_4,

and time rate of flow. The emerging structures obtained in this manner are termed as constructal designs. The same objective and constraints resulting in the similar structure that accommodates optimally shaped flow paths occurring in nature and artificial systems are named as constructal law [15–22]. It is the single theory encompassing the observations covered in animate and inanimate flow systems. In the pertinent literature [15], constructal law is synonymously recognized as the fourth law of thermodynamics. In constructal law, the competition between diffusion-like slower processes and convection-like faster processes are observed. Constructal law and the law of motive force are two self-standing independent laws. This chapter especially addresses to discover the commonalities between these two laws of nature.

It has long been observed that many of the volume-to-point and point-to-volume flows occurring in nature are in the form of tree networks [23–25]. The urge for formulating physics-based theory for the generating mechanism, from which fractal-like but not actually fractal rather purely Euclidean structure [26–28] could be predicted, was first met by the constructal theory of volume-to-point flows [16, 18–22].

The constructal theory was born of engineering optimization of paths of minimum thermal resistance for cooling finite-size small-scale electronic components [16]. The problem is to cool a finite-size volume by pure conduction. The statement of this fundamental problem is as follows. Consider a finite-size volume in which heat is being generated at every point and which is cooled through a small patch (heat sink) located on its boundary. A finite amount of high conductivity material is available. Determine the optimal distribution of such high conductivity material through the given volume such that the highest temperature is minimized [29]. The predicted structure reveals a manifestation of the principle of equipartition: the temperature drop through the high conductivity insert is equal to the temperature drop through low conductivity matrix [30]. The second important feature of this optimum can be recorded from the expression of minimized maximum temperature difference ΔT, which scales with the square of orthogonal dimension H of the heat conducting volume to the direction of applied heat current [30], i.e.,

$$\Delta T \sim H^2. \tag{4.1}$$

Thus, a power law correlates temperature drop and lateral dimension of the cooled volume. The message is to manufacture the smallest possible elemental system.

In another realistic access optimization problem, we arrive at a situation of point-to-volume flow. The statement of this fundamental problem is described below. Consider a fluid network to bathe a finite-size volume. The function of the path network is to distribute a stream of fluid to every elemental volume of the space. The mass flow rate of the fluid is purely due to pressure gradient (Hagen-Poiseuille) of the flow. The pressure differential varies with the position of the elemental volume relative to the point source. The maximum pressure difference, which is demanded by the elemental volumes, are those situated furthest from the source and are of specific importance. The total mass flow rate is fixed. The thermodynamic optimization of this fluid network is equivalent to minimizing the maximum pressure difference [31]. Optimized result yields that minimized maximum pressure drop ΔP scales with the square of the orthogonal dimension of the bathed volume to the direction of applied fluid flow [32], i.e.,

$$\Delta P \sim H^2. \tag{4.2}$$

The lesson of this power law correlation is to construct the narrowest possible elemental system. If the bifurcation of each path is assumed, each path width shrinks by a factor of $\frac{1}{2}$ from one stage to the next smaller stage [33]. Once again, the principle of equipartition is the crucial underlying feature.

The objective of this chapter is to discuss point-to-point, which we call elemental Fermat type flow, and volume-to volume, which we call integral Fermat type flow situations with reference to a fluid flow system. The results can be extended to a heat transporting system. The analogy and similarity between heat current and fluid stream is theoretically well established [34]. For further recognizing the qualitative similarities between heat and fluid flow it is important that we view them as a common entity of a flow field [35–37] as well as from the perspective of continuity [38]. In a similar fashion establishing a link between the law of motive force and the constructal law, we discover that in both the laws equipartition [39–43] is a common feature. The law of motive force is by virtue of its essence an expression for equipartition principle. The contrast between Fermat's principle and constructal law is well established and enunciated with clarity by Bejan [44]. The present state of the art of pattern formation study in fluid flow systems is generally accomplished through statistical theories [45, 46], numerical methods [47], and nonlinear dynamics [48]. Rigorous mathematical progress on the variational formulation for fluid flow problems [49–51] has also been made, but with little achievement in obtaining the shape and structure of the flow. The point of the present investigation is that the simple dimensional analysis [52–54] and scaling analysis [55–57] can be much revealing in capturing the physics of the problem. The message of this analysis is to echo Bejan: it is too early to give up on pencil and paper [58]. This current contribution elaborates the result obtained by the author [43] in view of the proposed law of motive force in this monograph.

4.2 Elemental Fermat Type Flow

For a large number of classes of naturally organized (self-organized) systems, it is important to establish the effect of gravitation on the thermodynamic properties of the systems [59]. First, it is instructive to establish the distribution of pressure p and specific volume v along the height h of a stack of fluid column. Then we must use an empirical equation of state for the given substance in the functional form $v = v(p)$, else we must use the method of successive approximations for which we need either experimental data or values calculated via the equation of state both relating to the p versus v dependence along the specific isotherm for the substance studied.

From the basic hydrostatic law, it is known that in a column of fluid the pressure varies with height. The change in pressure along the elementary column of height $\mathrm{d}h$ is

$$\mathrm{d}p = -\frac{\gamma}{A}\mathrm{d}V \tag{4.3}$$

where γ is the specific weight of the fluid in the column, $\mathrm{d}V$ is the elementary volume, and A is the cross-sectional area of the elementary column. Since $\mathrm{d}V = A\mathrm{d}h$, Eq. (4.3) reduces to

$$\mathrm{d}p = -\gamma \mathrm{d}h. \tag{4.4}$$

By definition $\gamma = \frac{g}{v}$ where g is the gravitational acceleration. Thus, we arrive at the following equation:

$$\mathrm{d}p = -\frac{g}{v}\mathrm{d}h \tag{4.5}$$

where the minus sign shows that with increasing height ($\mathrm{d}h > 0$) the fluid pressure decreases ($\mathrm{d}p < 0$). With the choice of reference frame at the top of the free surface instead of the bottom, this sign convention is reversed.

If the pressure p and temperature T of the gas are such that the fluid can be regarded as ideal, the equation of state translates into

$$v = \frac{RT}{p} \tag{4.6}$$

where R is the universal gas constant. In view of Eq. (4.6) we can rewrite Eq. (4.5) as

$$\mathrm{d}p = -\frac{p}{RT}\mathrm{d}h \tag{4.7}$$

whence

$$\frac{\mathrm{d}p}{p} = -\frac{\mathrm{d}h}{RT}. \tag{4.8}$$

Integrating Eq. (4.8) with respect to a reference pressure p_1 at a reference height h_1 we obtain

$$\ln\frac{p(h)}{p_1} = -\frac{1}{R}\int_{h_1}^{h}\frac{\mathrm{d}h}{T}. \tag{4.9}$$

For isothermal fluid column, we have

$$\ln\frac{p(h)}{p_1} = -\frac{h - h_1}{RT}. \tag{4.10}$$

Thus, we obtain the following formula for the distribution of pressure in an ideal gas isothermal column known as barometric height formula:

$$p(h) = p_1 \exp\left(-\frac{h - h_1}{RT}\right). \tag{4.11}$$

Invoking the ideal gas law (4.6) into Eq. (4.11) we find that

$$v(h) = v_1 \exp\left(\frac{h - h_1}{RT}\right) \tag{4.12}$$

where v_1 is the reference specific volume at reference pressure p_1. Hence, we see from the last but one relationship (4.11), dependence of pressure on height of the fluid column is of exponential nature. For small argument of the exponent, the relationship is almost linear.

In view of constructal theory as well as the law of motive force in a self-organized or engineered system, certain entities are equipartitioned. For a point-to-point flow configuration, distribution of pressure is of concern. We are interested to learn how the height H of an isothermal vertical fluid column can be divided into n horizontal parts so that pressure is equal in each subdivision.

Let w be the width of the fluid column. Suppose, below the top of the fluid column h_1 and h_2 are the depths of the two horizontal lines that divide the column into three portions. Say p_1, p_2, and p_3 are the three pressures, respectively, from the surface of fluid on the three portions of the column. The expressions for pressure can be written as follows:

$$p_1 = \frac{1}{2}\gamma w h_1^2, \tag{4.13}$$

$$p_2 = \frac{1}{2}\gamma w\left(h_2^2 - h_1^2\right), \tag{4.14}$$

and

$$p_3 = \frac{1}{2}\gamma w\left(H^2 - h_2^2\right). \tag{4.15}$$

Now, we impose the condition

$$p_1 = p_2 = p_3. \tag{4.16}$$

Eliminating the pressure term between Eqs. (4.13) and (4.14) we have

$$h_1 = \left(\frac{1}{2}\right)^{1/2} h_2. \tag{4.17}$$

Again, eliminating the pressure term between Eqs. (4.14) and (4.15) we arrive at

$$h_2 = \left(\frac{1}{2}\right)^{1/2} \left(h_1^2 + H^2\right)^{1/2}. \tag{4.18}$$

Solving Eqs. (4.17) and (4.18) for h_1 and h_2 in terms of H we obtain

$$h_1 = \left(\frac{1}{3}\right)^{1/2} H \tag{4.19}$$

and

$$h_2 = \left(\frac{2}{3}\right)^{1/2} H. \tag{4.20}$$

Following the method of induction [60] in general, we can write

$$h_i = \left(\frac{i}{n}\right)^{1/2} H \tag{4.21}$$

for $i = 1, 2, 3, \ldots, n$.

The center of pressure can be determined to find the coordinate of a representative pressure differential as the role played by center of mass in solid mechanics in place of a rigid body. Let $\bar{h}_1$, $\bar{h}_2$, and $\bar{h}_3$ be the depth of center of pressures below the top surface of the fluid column for the three portions of the column. The location of hydrostatic force $\bar{h}_P$ with respect to some pole P is related to the location of hydrostatic force $\bar{h}$ with reference to centroid C by the parallel-axis theorem [61] as follows:

$$\bar{h}_p = \bar{h} + I_C \frac{\sin^2 \alpha}{A\bar{h}} \tag{4.22}$$

where I_C is the moment of inertia with respect to the centroid and α is the inclination of the fluid column with the horizontal. Here, in particular

$$\alpha = \frac{\pi}{2},\ I_C = \frac{1}{12} wh^3,\ A = wh, \text{ and } \bar{h} = \frac{h}{2}.$$

Thus, for the first partition from the top, we have

$$\bar{h}_1 = \frac{2}{3} h_1. \tag{4.23}$$

Substituting back the value from Eqs. (4.19) into (4.23) we arrive at

$$\bar{h}_1 = \frac{2}{3}\left(\frac{1}{3}\right)^{1/2} H. \tag{4.24}$$

Similarly, for the second partition from the top we get

$$\bar{h}_2 = \frac{2}{3}\left(\frac{2^{3/2} - 1}{3^{1/2}}\right)H \tag{4.25}$$

and for the third portion from the top we obtain

$$\bar{h}_3 = \frac{2}{3}\left(\frac{3^{3/2} - 2^{3/2}}{3^{1/2}}\right)H. \tag{4.26}$$

Thus, generalizing the result on following the method of induction [60] we finally arrive at

$$\bar{h}_i = \frac{2}{3}\left(\frac{i^{3/2} - (i-1)^{3/2}}{n^{1/2}}\right)H \text{ for } i = 1, 2, 3, \ldots, n. \tag{4.27}$$

Contrary to the equipartition of the physical quantity pressure, we now choose to consider equipartition of space and then to seek the distribution of pressure therein. Let the fluid column H be divided into n large number of equal-sized slices such that

$$h = \frac{H}{n}. \tag{4.28}$$

Suppose the densities of these layers are ρ_1, ρ_2, $\rho_3, \ldots$, ρ_n, respectively. These densities are practically constant over these small slices. Obeying ideal gas equation of state, their corresponding pressures are $RT\rho_1$, $RT\rho_2$, $RT\rho_3$, and $RT\rho_n$, respectively. Since, the size of the slices is small, the same pressure is valid at all points of the slice. It means that center of pressure is of no specific importance here. Again, in the infinitesimal sense, the difference in pressures on the top and bottom faces of a slice is equal to the weight of the fluid contained in the layer. Hence, we can write in succession

$$RT\rho_1 - RT\rho_2 = \rho_1 gh, \tag{4.29a}$$

$$RT\rho_2 - RT\rho_3 = \rho_2 gh, \tag{4.29b}$$

and

$$RT\rho_{n-1} - RT\rho_n = \rho_{n-1} gh. \tag{4.29c}$$

From Eq. (4.29a), we get

$$\rho_2 = \rho_1\left(1 - \frac{gh}{RT}\right). \tag{4.30a}$$

Similarly, from Eq. (4.29b) using the result of Eq. (4.29a) we obtain

$$\rho_3 = \rho_2\left(1 - \frac{gh}{RT}\right) = \rho_1\left(1 - \frac{gh}{RT}\right)^2. \tag{4.30b}$$

Thus, in general we can write

$$\rho_n = \rho_{n-1}\left(1 - \frac{gh}{RT}\right) = \rho_1\left(1 - \frac{gh}{RT}\right)^{n-1}. \tag{4.30c}$$

Hence, as the altitude increases in arithmetic progression, the densities and the corresponding pressures decrease in geometric progression from the bottom of the vertical fluid column. Now, if ρ is the density just above the nth layer, from Eq. (4.30c) we have

$$\rho = \rho_n\left(1 - \frac{gh}{RT}\right) = \rho_1\left(1 - \frac{gh}{RT}\right)^{n}. \tag{4.30d}$$

Invoking Eq. (4.28) into Eq. (4.30d) and rewriting we arrive at the expression

$$\rho = \rho_1\left[\left(1 - \frac{1}{z}\right)^{-z}\right]^{\left(-\frac{gH}{RT}\right)} \tag{4.30e}$$

where

$$z = \frac{nRT}{gH}.$$

For $n \to \infty$, H remains constant but $z \to \infty$. Recognizing the limit [62]

$$\lim_{z\to\infty}\left(1 - \frac{1}{z}\right)^{-z} = e \tag{4.30f}$$

we finally have

$$\rho = \rho_1 \exp\left(-\frac{gH}{RT}\right). \tag{4.30g}$$

Following the ideal gas law, the expression for pressure takes on the form

$$p = p_1 \exp\left(-\frac{gH}{RT}\right). \tag{4.30h}$$

As expected, the Eq. (4.30h) is identically the same as that of Eq. (4.11) obtained earlier. The message of the above analysis is that equipartition of one entity demands the power law distribution of the other associated with it.

Next, we may be interested to learn the range of values of the index of the power law distribution of a physical quantity for which the equipartition of the other quantity is valid. In the following example, we consider the gauge pressure distribution on the face of a vertical rectangular sluice gate in a free surface flow.

From the experimental evidence, the gauge pressure distribution conforms to a mathematical relation of the form [63]

$$p - p_{\text{atm}} = \rho g h \left[1 - \left(\frac{h}{H} \right)^{n} \right] \tag{4.31}$$

where p_{atm} is the atmospheric pressure exerted on the free surface of the flow, H is the depth of the gate, and n is a parametric constant. We are interested to estimate the magnitude and location of the resulting horizontal force on the gate.

Elemental pressure force $\mathrm{d}F_x$ in the horizontal direction on an elemental strip of width w and height $\mathrm{d}h$ is

$$\mathrm{d}F_x = (p - p_{\text{atm}}) w \mathrm{d}h. \tag{4.32}$$

Using Eq. (4.31) into Eq. (4.32) we get

$$\mathrm{d}F_x = \rho g w \left(h - \frac{h^{n+1}}{H^n} \right) \mathrm{d}h. \tag{4.33}$$

Total horizontal force is obtained upon integrating Eq. (4.33) as

$$F_x = \frac{1}{2} \rho g w H^2 \left(\frac{n}{n+2} \right). \tag{4.34}$$

Employing the concept of averaging, we calculate the location h_P of hydrostatic force as

$$h_p F_x = \int_0^H h \mathrm{d}F_x. \tag{4.35}$$

Substituting the expressions for $\mathrm{d}F_x$ and F_x from Eqs. (4.33) and (4.34), respectively, we get

$$h_p = \frac{2}{3} H \left(\frac{n+2}{n+3} \right). \tag{4.36}$$

On passing to the limit $n \to \infty$ in Eqs. (4.34) and (4.36) we obtain, respectively,

$$\lim_{n \to \infty} (F_x) = \frac{1}{2} \rho g w H^2 \tag{4.37}$$

and

$$\lim_{n \to \infty} (h_p) = \frac{2}{3} H. \tag{4.38}$$

Thus, Eqs. (4.37) and (4.38) are asymptotic to the usual results when the index of the power law is very great.

4.3 Integral Fermat Type Flow

It is imperative to be curious about the happenings around us to learn the functioning mechanism of nature. An example of such cadre is the hydraulic jump, which often takes place in the study of river morphology. It is a sudden discontinuity in the depth of the flowing fluid. During the period of tide, a jump may sometimes be observed by standing or moving upstream. This phenomenon of jump can easily be reproduced in laboratory scale. A plate held horizontally under the faucet of fluid may be employed to demonstrate a hydraulic jump. The moving fluid is allowed to hit the center of the plate. Then the fluid flows radially outward in the form of a fast thin layer and suddenly increases in thickness before flowing over the edge of the plate. We are interested in examining the relationship between the upstream and downstream thickness responsible for the mechanism of elbow growth and eddy formation in terms of relevant parameters to testify the validity of certain power laws and the equipartition principle.

Let us consider a control volume of width w with the paper. Thus, the jump can be treated as stationary with respect to the control volume. Assume the velocities V_1 before jump and V_2 after jump are uniform over the channel. By choosing the control volume to be very thin, the frictional force on the channel bed may be neglected. Let h_1 and h_2 be the heights of the fluid stream before and after jump, respectively. For a bulk flow model [64], the density ρ may be considered constant for a small volumetric discharge Q through the control volume. Applying continuity equation we have

$$\rho w h_1 V_1 = \rho w h_2 V_2 = Q. \tag{4.39}$$

Hydrostatic pressure forces over each face of control volume can be accounted for momentum transfer across the faces and thus we get

$$\frac{1}{2} g h_1^2 - \frac{1}{2} g h_2^2 = V_2^2 h_2 - V_1^2 h_1. \tag{4.40}$$

In Eq. (4.40), the mechanism of momentum transport resembles that of a conservation principle pertaining to the competition between backward and forward motivation. The forward motivation can be treated as due to velocity such as $V_1^2 h_1$ and $V_2^2 h_1$. The backward motivation can be counted as due to gravitational acceleration such as $\frac{1}{2} g h_1^2$ and $\frac{1}{2} g h_2^2$. A simple dimensional analysis will further confirm the fact. From Eq. (4.39), we obtain

$$V_2 = \left(\frac{h_1}{h_2}\right) V_1. \tag{4.41}$$

Invoking Eq. (4.41) into Eq. (4.40) we formulate a quadratic equation in h_1 and h_2. Trivial solution of this equation leads to

$$h_1 = h_2. \tag{4.42}$$

Nontrivial solution is to be extracted from the following expression:

$$\left(\frac{h_2}{h_1}\right)^2 + \left(\frac{h_2}{h_1}\right) - 2\left(\frac{1}{g}\right)\left(\frac{V_1^2}{h_1}\right) = 0. \tag{4.43}$$

Invoking $V_1 = \frac{Q}{\rho w h_1}$ from Eq. (4.39) into Eq. (4.43), we get

$$\frac{h_2}{h_1} = -\frac{1}{2} + \frac{1}{2}\sqrt{1 + (2)^2\left(\frac{1}{g\rho^2}\right)\left(\frac{Q^2}{w^2 h_1^3}\right)}. \tag{4.44}$$

For $h_2 \sim h_1 \sim h$ (say) we must have

$$(2)^2\left(\frac{1}{g\rho^2}\right)\left(\frac{Q^2}{w^2 h_1^3}\right) \sim 0. \tag{4.45}$$

It implies that

$$h_1 \gg (2)^{2/3}\left(\frac{1}{g}\right)^{1/3}\left(\frac{Q}{\rho w}\right)^{2/3}. \tag{4.46}$$

This means h_1 has a minimum of the following order:

$$h_{1,\min} \sim (2)^{2/3}\left(\frac{1}{g}\right)^{1/3}\left(\frac{Q}{\rho w}\right)^{2/3}. \tag{4.47}$$

Thus, it can be concluded that for a fluid with definite flow geometry, depth of flow h_2 after jump scales with $\frac{2}{3}$ power of the stream volume, i.e.,

$$h_{2,\min} \sim Q^{2/3}. \tag{4.48}$$

Next, it is interesting to recognize the results as h_1 approaches h_2, the jump becomes a small surface wave. From the energy considerations and the second law of thermodynamics, we confirm the fact that $V_2 < V_1$ and $h_2 > h_1$, as energy must be lost by friction through the jump. This Cauchy-Poisson problem [65] of small amplitude wave was studied theoretically by Rayleigh [66], Kochin [67], and Sedov [68]. Reynolds [69] performed an experimental investigation. Recently, Bejan [70] showed that the general solution of such small-amplitude wavelength $h(x)$ in the longitudinal direction x scales with a sinusoidal function of the form

$$h(x) \sim \sin^2\left(\frac{1}{2}x\sqrt{\frac{I}{A}}\right) \tag{4.49}$$

where A is the cross-sectional area and I is the area moment of inertia of the stream.

As the height difference is not appreciable before and after the jump, the energy is conserved on following the law of motive force. In view of Eq. (4.49), flow energy is equipartitioned in the post-buckled (degenerated) stream between the upper and lower halves of a sinusoid.

Now, we calculate the velocity of propagation of this small-amplitude wave. From Eq. (4.39) in view of negligible jump we obtain

$$V_1 \sim V_2 \sim V \text{ (say)}. \tag{4.50}$$

Invoking Eq. (4.50) into Eq. (4.43) we obtain

$$V = \sqrt{gh}. \tag{4.51}$$

Rearranging Eq. (4.51) in the form

$$V = \left(\frac{1}{2}\right)^{1/2} \sqrt{2gh} \tag{4.52}$$

we see that velocity of flow for a negligible hydraulic jump is a scale factor $\left(\frac{1}{2}\right)^{1/2}$ of the efflux from a narrow opening at the bottom of the stream.

The maximum amplitude of the elbow is of the order [71] of $\frac{h}{2}$ and this result is confirmed by all observations of free jet flows. The post-buckled elbow region becomes a distinct eddy. If the stream (h, V) was already carrying small eddies, a large-scale turbulent structure continues to move downstream with a speed [72] of the order of $\frac{V}{2}$. This is also an instance of equipartition of velocity.

4.4 First Geometrical Construct in a Shear Flow

From the discussion of elemental Fermat type flow, it is evident that a stable fluid column can exist in the form of a vertical and/or horizontal line segment in one-dimensional arrangement. Thus, the most natural choice of a fluid element in a two-dimensional static situation is in the form of a finite rectangular block. In the flow situation, the geometry assumes the shape of a parallelogram. It can be guaranteed that the smallest angle θ, measured in radians and counterclockwise positive, between the two nonparallel sides of the configuration is bounded in the domain $0 \leq \theta \leq \frac{\pi}{2}$.

We consider an identified elemental area of the flow field in the form of a parallelogram $ABCD$ as in Fig. 4.1a. It is exposed to fluid pressure due to its self-weight and the force exerted by the adjacent layers in a flow situation. Its sides AB and AD are x and y, respectively.

We are interested to recognize the basic geometrical shape of fluid element responsible for pressure and kinetic energy transport in a flow. We also examine the validity of the continuum principle at every point of the flow. Figure 4.1b presents an exaggerated view of the parallelogram CC'.

Let, the thrust on the area $ABCD$ be $F(x, y)$, which is a continuous function in space variable. We complete the parallelogram $AB'C'D'$ with sides $(x + \delta x)$ and $(y + \delta y)$. Area of the elementary parallelogram CC' is $\delta x \delta y \sin\theta$. Thrust on area CC' can be expressed as

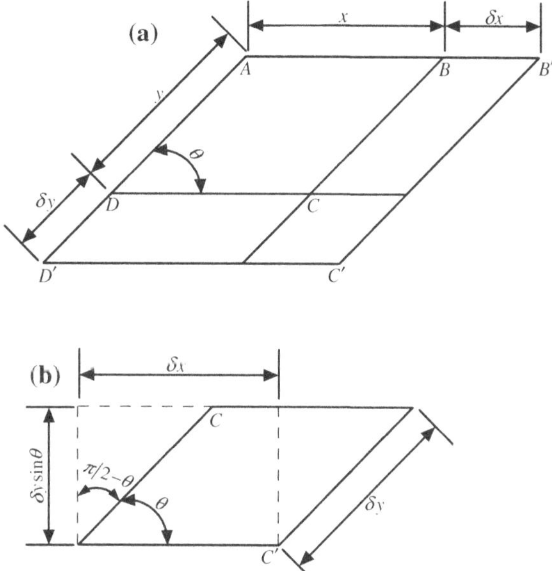

Fig. 4.1 **a** First construct in a shear flow; **b** Exaggerated view of an elemental shear transporting block CC'

$$F(x,y)|_{CC'} = F(x+\delta x, y+\delta y) - F(x+\delta x, y) - F(x, y+\delta y) + F(x,y). \quad (4.53)$$

Then the pressure on CC' defined as thrust per unit area, appears as

$$p|_{CC'} = \frac{1}{\sin\theta}\lim_{\delta x\to 0}\left[\frac{\lim\limits_{\delta y\to 0}\frac{F(x+\delta x,y+\delta y)-F(x+\delta x,y)}{\delta y} - \lim\limits_{\delta y\to 0}\frac{F(x,y+\delta y)-F(x,y)}{\delta y}}{\delta x}\right]. \quad (4.54)$$

Performing the sum of limits, the expression for pressure becomes

$$p|_{CC'} = \left(\frac{1}{\sin\theta}\right)\frac{\partial^2 F}{\partial x \partial y}. \quad (4.55)$$

If the limits were performed in a different order, we would obtain

$$p|_{CC'} = \left(\frac{1}{\sin\theta}\right)\frac{\partial^2 F}{\partial y \partial x}. \quad (4.56)$$

Since, the thrust $F(x, y)$ is continuous in space variable, we have from Eqs. (4.55) and (4.56) the uniqueness of pressure as

$$p|_{CC'} = \left(\frac{1}{\sin\theta}\right)\frac{\partial^2 F}{\partial x \partial y} = \left(\frac{1}{\sin\theta}\right)\frac{\partial^2 F}{\partial y \partial x}. \quad (4.57)$$

Thus, the pressure is continuous even at the corner point where it could be singular. Further, it is to be noted that the load bearing capacity of the fluid element is maximum when $\theta = \frac{\pi}{2}$ and it is undefined for a horizontal line element when $\theta = 0$. It is interesting to report that maximum shear transport occurs when $\theta = \frac{\pi}{4}$ which is the mean value of the upper and lower bounds of the included angle. If we set now $\theta = \theta_1$ and $\frac{\pi}{2} - \theta = \theta_2$ the essence of flow configuration may be represented as $\theta_1 + \theta_2 = \frac{\pi}{2}$ (constant). The order of magnitude of θ_1 is an indication of the flow strength and can be regarded as forward motivation. On the other hand, depending on the shear resistance of the flow the value of angle θ_2 is automatically adjusted resulting in the shape of the flow geometry. Hence, angle θ_2 can be treated as a backward motivation for this flow situation.

Next, from the flow configuration employing transformation geometry it can be conceived that rectangular shape is altered by cutting a triangular slice from the left-hand side and translating it to the right-hand side for a pressure transmission from left to right. For a finite-size system, the elemental block could be considered comparably small to conceptualize that the shear flow takes place essentially in the form of tiny wedge packets. For pure Couette type (shear driven) flow the interpretation is obvious. For Hagen-Poiseullie (pressure driven) flow, the situation can be thought of as two superimposed Couette flows with a moving boundary at the mean line of the flow geometry. The idea expressed here can be easily extended to the local potential model [73] and the stability [74–77] problems of Couette flow [78].

4.5 Discussions

Understanding the physics of the problem can greatly simplify the mathematical calculation process of thermodynamic optimization of systems. Post analysis of the results obtained by virtue of constructal principle exhibits the property of equipartition of entities between two potentially competing forces of forward and backward motivation. Thus, the optimization of a thermal system, in specific, is instructive as follows. First, to choose all factors affecting system performance. An orders of magnitude analysis is to be invited to eliminate the factors not of significant contributions. In all the situations two categories of competing forces result. They can be added up to a constant to obtain the optimum values of the parameter in concern. This design procedure is as rigorous as other optimization methodologies in any purely mathematical prescription. The physical reality is to cast higher order constructs from smaller ones by what we implement as method of induction.

Another characteristic feature of the law of motive force and constructal theory is to predict finite shape, which is featured in the application of finite-time thermodynamics. An argument from the Euclidean geometrical frame is established to predict geometric form for the first construct in a shear flow. The finding that flow proceeds in the form of wedge packet is at par with the stipulation of continuum mechanics and also conforms with the observation covered in other natural flow processes such as flow of radiation, information, etc.

References

1. Bejan, A.: Entropy Generation Minimization. CRC, Boca Raton (1996)
2. Bejan, A.: Notes on the history of the method of entropy generation minimization (finite time thermodynamics). J. Non-Equilib. Thermodyn. **21**, 239–242 (1996)
3. Szargut, J., Morris, D.R., Steward, F.R.: Exergy Analysis of Thermal, Chemical and Metallurgical Processes. Hemisphere, New York (1988)
4. Yantovskii, E.I.: Energy and Exergy Currents. Nova Science, New York (1994)
5. Chen, L., Sun, F. (eds.): Advances in Finite Time Thermodynamics: Analysis and Optimization. Nova Science, New York (2004)
6. Sieniutycz, S., Salamon, P. (eds.): Finite-Time Thermodynamics and Thermoeconomics. Taylor & Francis, New York (1990)
7. Chambadal, P.: Les Centrales Nucleaires, pp. 41–58. Armand Colin, Paris (1957) (in French)
8. Landsberg, P.T., Leff, H.S.: Thermodynamic cycles with nearly universal maximum-work efficiencies. J. Phys. A Math. Gen. **22**, 4019–4026 (1989)
9. Gaggioli, R. (ed.): Efficiency and Costing. ACS Symposium Series, vol. 235. ACS, Washington (1983)
10. Tribus, M., Evans, R.: The thermoeconomics of sea water conversion. UCLA Report No. 62-63, August 1962
11. Rozonoer, L.I., Tsirlin, A.M.: Optimal control of thermodynamic processes-I. Autom. Remote Control **44**, 55–62 (1983)
12. Rozonoer, L.I., Tsirlin, A.M.: Optimal control of thermodynamic processes-II. Autom. Remote Control **44**, 209–220 (1983)
13. Rozonoer, L.I., Tsirlin, A.M.: Optimal control of thermodynamic processes-III. Autom. Remote Control **44**, 314–326 (1983)
14. Salamon, P., Nulton, J.D., Siragusa, G., Andresen, T.R., Limon, A.: Principles of control thermodynamics. Energy **26**, 307–319 (2001)
15. Bejan, A.: Advanced Engineering Thermodynamics, p. 807. Wiley, New York (1997)
16. Bejan, A.: Constructal-theory network of conducting paths for cooling a heat generating volume. Int. J. Heat Mass Transf. **40**, 799–816 (1997)
17. Bejan, A.: Shape and Structure, from Engineering to Nature, pp. 60–62. Cambridge University Press, Cambridge (2000)
18. Bejan, A., Dan, N.: Two constructal routes to minimal heat flow resistance via greater internal complexity. J. Heat Transf. **121**, 6–14 (1999)
19. Bejan, A., Errera, M.R.: Deterministic tree networks for fluid flow: geometry of minimum flow resistance between a volume and one point. Fractals **5**, 685–695 (1997)
20. Bejan, A., Ledezma, G.A.: Streets tree networks and urban growth: optimal geometry for quickest access between finite-size volume and one point. Physica A **255**, 211–217 (1998)
21. Dan, N., Bejan, A.: Constructal tree networks for the time dependent discharge of a finite-size volume to one point. J. Appl. Phys. **84**, 3042–3050 (1998)
22. Ledezma, G.A., Bejan, A., Errera, M.R.: Constructal tree networks for heat transfer. J. Appl. Phys. **82**, 89–100 (1997)
23. Crammer, F.: Chaos and Order. VCH, Weinheim (1993)
24. Prigogine, I.: From Being to Becoming. Freeman, New York (1980)
25. Thompson, D.W.: On Growth and Form. Cambridge University Press, Cambridge (1942)
26. Avnir, D., Biham, O., Lidar, D., Malacai, O.: Is the geometry of nature fractal? Science **279**, 39–40 (1998)
27. Bejan, A.: Advanced Engineering Thermodynamics, pp. 739–742. Wiley, New York (1997)
28. Nottale, L.: Fractal Space-Time and Microphysics. World Scientific, Philadelphia (1982)
29. Bejan, A.: Advanced Engineering Thermodynamics, p. 724. Wiley, New York (1997)
30. Bejan, A.: Advanced Engineering Thermodynamics, p. 727. Wiley, New York (1997)
31. Bejan, A.: Advanced Engineering Thermodynamics, p. 745. Wiley, New York (1997)
32. Bejan, A.: Advanced Engineering Thermodynamics, p. 749. Wiley, New York (1997)

33. Bejan, A.: Advanced Engineering Thermodynamics, p. 743. Wiley, New York (1997)
34. Bejan, A.: Shape and Structure, from Engineering to Nature. Cambridge University Press, Cambridge (2000)
35. Parkus, H., Sedov, L.I. (eds.): Irreversible Aspect of Continuum Mechanics and Transfer of Physical Characteristics in Moving Fluids. Springer, Berlin (1968)
36. Sedov, L.I.: Introduction to the Mechanics of Continuous Medium. Addison-Wesley, New York (1965)
37. Sedov, L.I. (ed.): Macroscopic Theories of Matter and Fields: A Thermodynamic Approach (trans: Yankovsky, E.), pp. 19–42, 43–97. Mir, Moscow (1983)
38. Van Der Waals, J.D.: On the Continuity of Gaseous and Liquid States. In: Rowlinson, J.S. (ed.). Dover, New York (2004)
39. Bejan, A.: Advanced Engineering Thermodynamics, pp. 352–356, 464–466, 569–571, 709–721, 782–788, 816–820. Wiley, New York (2006)
40. Bejan, A.: Shape and Structure, from Engineering to Nature, pp. 53–56, 84–88, 99–108, 151–161, 220–223, 234–242, 287–288. Cambridge University Press, Cambridge (2000)
41. Bejan, A., Tondeur, D.: Equipartition, optimal allocation, and the constructal approach to predicting organization in nature. Rev. Gen. Therm. **37**, 165–180 (1998)
42. De Vos, A., Desoete, B.: Equipartition principle in finite-time thermodynamics. J. Non-Equilib. Thermodyn. **25**, 1–13 (2000)
43. Pramanick, A.K., Das, P.K.: Note on constructal theory of organization in nature. Int. J. Heat Mass Transf. **48**, 1974–1981 (2005)
44. Bejan, A.: Constructal comment on a Fermat-type principle for heat flow. Int. J. Heat Mass Transf. **46**, 1885–1886 (2003)
45. Monin, A.S., Yaglom, A.M.: Statistical Fluid Mechanics-I. In: Lumley, J.L. (ed.). Dover, New York (2007)
46. Monin, A.S., Yaglom, A.M.: Statistical Fluid Mechanics-II. In: Lumley, J.L. (ed.). Dover, New York (2007)
47. Yanenko, N.N., Shokin, Yu.I. (eds.): Numerical Methods in Fluid Dynamics (trans: Shokurov, V., Hainsworth, R.N. (ed.)). Mir, Moscow (1984)
48. Sagdeev, R.Z.: Nonlinear Phenomena in Plasma Physics and Hydrodynamics (trans: Ilyushchenko, V.). Mir, Moscow (1986)
49. Bergman, S., Schiffer, M.: Kernel Functions and Elliptic Differential Equations in Mathematical Physics, pp. 59–64. Dover, New York (2005)
50. Friedman, A.: Variational Principles and Free-Boundary Problems. Dover, New York (2010)
51. Lavrent'ev, M.A.: Variational Methods for Boundary Value Problems for Systems of Elliptic Equations (trans: Radok, J.R.M.), pp. 42–71. Dover, New York (2006)
52. Hornung, H.G.: Dimensional Analysis. Dover, New York (2006)
53. Sedov, L.I.: Similarity and Dimensional Methods in Mechanics (trans: Kisin, V.I.). Mir, Moscow (1982)
54. Yarin, L.P.: The Pi-Theorem. Springer, New York (2012)
55. Bejan, A.: Convection Heat Transfer, pp. 19–23. Wiley, New York (2004)
56. Bejan, A.: The method of scale analysis: natural convection in fluids. In: Kakac, S., Aung, W., Viskanta, R. (eds.) Natural Convection: Fundamentals and Applications. Hemisphere, Washington (1985)
57. Bejan, A.: The method of scale analysis: natural convection in porous media. In: Kakac, S., Aung, W., Viskanta, R. (eds.) Natural Convection: Fundamentals and Applications. Hemisphere, Washington (1985)
58. Bejan, A.: Advanced Engineering Thermodynamics, p. 811. Wiley, New York (1997)
59. Sychev, V.V.: Complex Thermodynamic Systems (trans: Yankovsky, E.), pp. 165–182. Mir, Moscow (1981)
60. Courant, R., Robbins, H.: What is Mathematics? (Stewart, I., Revised), pp. 9–20. Oxford University Press, Oxford (2007)
61. Vardy, A.: Fluid Principles, pp. 57–58. McGraw-Hill, New York (1990)

62. Courant, R., Robbins, H.: What is Mathematics? (Stewart, I., Revised), p. 478. Oxford University Press, Oxford (2007)
63. Vardy, A.: Fluid Principles, p. 73. McGraw-Hill, New York (1990)
64. Bejan, A.: Advanced Engineering Thermodynamics, p. 70. Wiley, New York (1997)
65. Lamb, H.: Hydrodynamics, pp. 384–394. Cambridge University Press, Cambridge (1974)
66. Rayleigh, L.: On the theory of long waves and bores. Proc. R. Soc. Lond. A **90**, 324–328 (1914)
67. Kochin, N.E.: On the theory of Cauchy-Poisson waves. Tr. MIAN SSSR **9** (1935) (in Russian)
68. Sedov, L.I.: On the theory of small-amplitude waves on the surface of an incompressible fluid. Vestnik MGU **11**, 71–77 (1948) (in Russian)
69. Reynolds, O.: An experimental investigation of the circumstances which determine whether the motion of water shall be direct or sinuous, and of the law of resistance in parallel channels. Philos. Trans. R. Soc. Lond. **174**, 935–982 (1883)
70. Bejan, A.: On the buckling property of inviscid jets and the origin of turbulence. Lett. Heat Mass Transf. **8**, 187–194 (1981)
71. Bejan, A.: Entropy Generation Through Heat and Fluid Flow, p. 75. Wiley, New York (1982)
72. Crow, S.C., Champagne, F.H.: Orderly structure in jet turbulence. J. Fluid Mech. **48**, 547–591 (1971)
73. Glansdorff, P., Prigogine, I.: Thermodynamic Theory of Structure, Stability and Fluctuations, pp. 126–153. Wiley-Interscience, New York (1971)
74. Heisenberg, W.: Uber Stabilitat und Turbulenz von Flussigkeitsstromen. Ann. Phy. Lpz. **74**, 577–627 (1924) (in German). Also NACA TM-1291 (1951)
75. Lin, C.-C.: On the stability of two-dimensional parallel flows. Proc. NAS **30**, 316–324 (1944)
76. Rayleigh, L.: On the stability, or instability of certain fluid motions. Proc. Lond. Math. Soc. **11**, 57–70 (1880)
77. Tollmien, W.: Ein allgemeines Kriterum der Instabilitat laminarer Gesegwindigkeitsverteilungen. Nachr. Wissfachgruppe, Göttingen Math. Phys. **1**, 79–114 (1935) (in German). Also NACA TM-792 (1936)
78. Chandrasekhar, S.: Hydrodynamic and Hydromagnetic Stability, pp. 272–342. Dover, New York (1981)

Chapter 5
Natural Heat Engine

> *The principles of thermodynamics occupy a special place among the laws of Nature. For this there are two reasons: in the first place, their validity is subject only to limitations which, though not, perhaps themselves negligibly small, are at any rate minimal in comparison with many other laws of Nature; and in the second place, there is no natural process to which they cannot be applied.*
>
> W. Nerst

In this chapter, we study the thermoelectric generator from the perspective of a heat engine, which in turn falls into a class of thermal insulation systems. We employ the method of finite-time thermodynamics to take into account the essential features of a realistic heat engine. We directly look into the geometrical shape and structure of the building blocks of each thermoelectric module of the cascaded assembly that eventually causes a better global performance. Search for better geometry yields the ideal thermophysical properties of thermoelectric materials. Such a methodology of directly looking into the optimum shape and structure of the hardware components suggests also the most natural constraint to be employed in view of the law of motive force in the optimization method. Generally, in any optimization process we merely seek the conditions and do not investigate its consequences. In the present strategy, both the conditions are consequences sought with equal priority to discover the omnipresence of the law of motive force. The quest for the symmetry in geometrical construction has also appeared as a consequence of the analysis. Equipartition principle is found as a common bridge between the law of motive force and the constructal law. The final architecture of a cascaded assembly of thermoelectric modules exhibits a fractal-like but deterministic pattern (constructal) that can be constructed either from the largest to the smallest scale or in the reverse direction with the fundamental construct being symmetrical, which is the T-shaped region in space.

5.1 The Problem

In the history of science and engineering, thermoelectric phenomenon is old and prevalent [1–8]. Thermoelectric device was considered to verify the second law of thermodynamics [9] and to model the heat engine [10–12]. It is generally

A. K. Pramanick, *The Nature of Motive Force*, Heat and Mass Transfer,
DOI: 10.1007/978-3-642-54471-2_5,

postulated or observed [13, 14] that exactly half of the Joulean heat produced in a thermoelectric device arrives equally both at the hot and cold junction. In this chapter, some of the conditions and consequences of this equipartitioned Joulean heat are reported.

Thermoelectric generator is a useful and environment friendly device for direct energy conversion. Especially, the capacity of Peltier and Seebeck effect to dispense with the moving parts in the realm of energy transformation from heat to electricity and vice versa is more appealing in such devices. With the advent of semiconductor materials the efficiency of a thermoelectric generator can even be an alternative for conventional heat engines [5]. Another perspective of thermodynamic modeling of a thermoelectric generator, also recognized as natural heat engine, is that it includes all the crucial features of a real heat engine in a relatively simple way where closed-form expressions are obtained for the power versus efficiency characteristics [11]. Here, each generic source of irreversibility is identified and quantified in this process to draw a one to one correspondence between the conventional heat engine and the thermoelectric generator. Hence, the mathematical modeling of a simple thermoelectric generator can also replace the elaborate task of simulating an actual complex power plant, heat engine, or refrigerator.

Much effort has been bestowed in finite-time thermodynamics (FTT) [15–20] to model real heat engines. FTT modeling of thermoelectric generator presents a full-featured analysis of real engines. Since all heat engine models aim at providing a realistic margin for an improvement of actual systems, an analysis based on FTT figure-of-merit hints at a more practical assessment of a maximum attainable improvement in comparison to the margin based on Carnot efficiency. Thus, FTT modeling is a worthy endeavor. However, it is to be noted that FTT modeling does not stipulate the highest ceiling for efficiency, but only dictates the lower bound of the optimal efficiency of a heat engine performance affected by finite heat transfer rate irreversibility [21]. In practice, heat engines can operate between the two extreme limits: one is the reversible or maximum efficiency operation and the other is the irreversible or maximum power condition. However, in practical situations, the optimum design criterion is a compromise between the efficiency and power output. In terminologies of thermoeconomics [22], optimum operating point is a trade-off between the cost of fuel and the cost of installed hardware.

In an FTT model, generally, all possible irreversibilities are attributed only to the heat transport process external to the engine and not to the internal conversion of heat into power [23]. For an FTT model of a thermoelectric generator, external irreversiblities remain in series. Bypass heat leak [24] incorporated into the modeling is an additional shunt among other possible alternatives [25–27] that make the engine to operate irreversibly. In this study, bypass heat leak is identified as a major contribution to the measure of internal irreversibility. The conducting mechanical support, which is the locus of heat transfer across a finite temperature gap, is the geometrical path of irreversibility transport. The bypass heat leak phenomenon retains all the essential features of irreversibility of the engine and offers an elegant mathematical perspective for engine modeling. Though in a thermoelectric generator Joulean heating itself remains as an inherent source of irreversibility, bypass heat leak has normally higher orders of magnitude than internal irreversibility alone in the range of optimum engine performance.

The architecture of optimized flow system, in general, is a commonplace occurrence in engineering and nature. Solutions of many challenges have been unified under the single encompassing physics-based principle, the constructal law [28] that conceives that the geometry (shape and structure) is generated in pursuit of global performance subject to global constraints, in flow systems the geometry of which is free to vary. In this context, a thermoelectric device can be thought of as a flowing system through which heat and electrical current flows simultaneously. Calculation of efficiency of a thermoelectric device is reported in the open literature [29]. There, the conditions and consequences of heat transport between the heating and the cooling media and the junctions are not addressed. The objective of the present investigation aims at reporting finite-time irreversibility of heat transport mechanism, the distribution of Joulean heat into the hot and cold space and its consequences on optimal allocation of heat exchanger inventory, and finally to predict the geometrical shape and size of each individual module of a cascaded thermoelectric device. As pointed out in [30], different assumptions can lead to different results; the role of assumptions in describing the model are also stressed. The present focus explores the result obtained by the author [31, 32] in view of the proposed law of motive force in this treatise.

5.2 The Physical Model

To manifest the effect of simultaneous electric current and heat transfer irreversibilities on the thermal efficiency of a thermoelectric power generator, we consider the two-leg assembly of the basic components of a device as shown in Fig. 5.1. The hot junction is maintained at a high temperature level T_{HC} and it receives a net heat transfer rate $\dot{Q}_H$. Similarly, the cold end of the two-leg arrangement is held at constant temperature T_{LC} such that the net heat rejection rate by it is $\dot{Q}_L$. Ideally, the two legs are one-dimensional conductors along which x is directed from T_{HC} to T_{LC}. The potential difference generated due to Seebeck effect causes the flow of a total electrical current I through the total electrical resistance R of the elementary module of length L of a cascaded system.

The two legs n and p are generally chosen to be of dissimilar semiconductors or semimetals. In a conventional junction design, hot ends of the two legs are both electrically and thermally connected through a highly conductive material. Thermodynamically, this arrangement is equivalent to that of a simple design in which n- and p-legs are joined end-to-end. The lateral surfaces of both the legs are insulated electrically and thermally to prevent contact from each other.

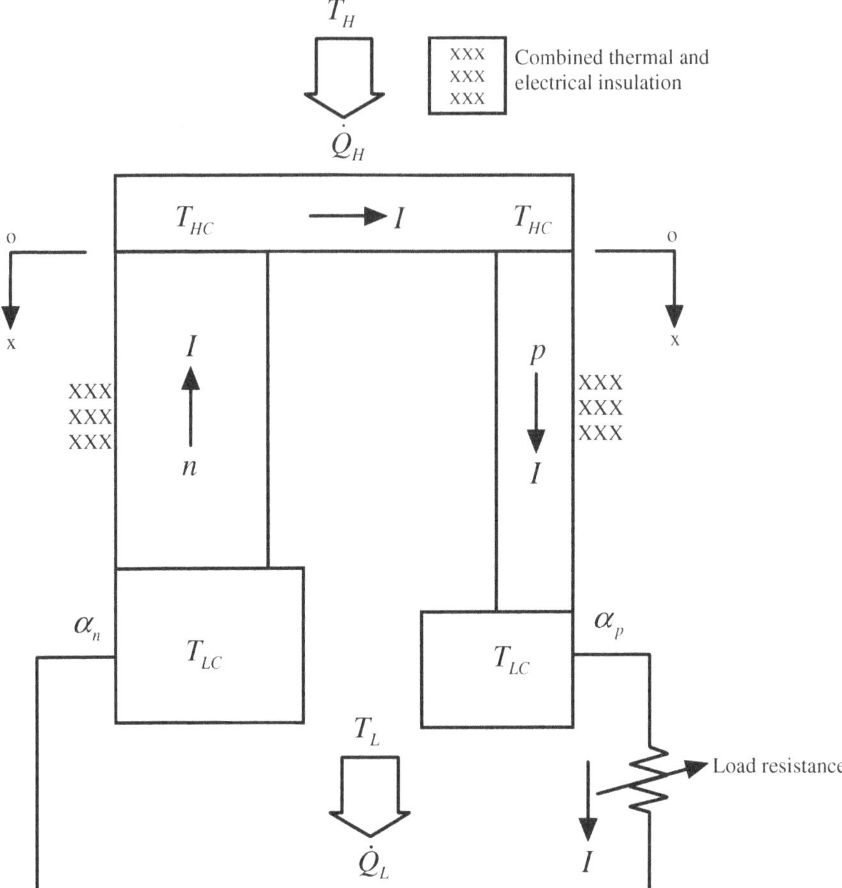

Fig. 5.1 Basic elements of a thermoelectric power generator

Additionally, the cold ends of the two legs are either insulated only electrically or situated separately from each other.

In the literature, thermodynamics of irreversible process is applied to a thermocouple where the legs may have an arbitrary shape and size, the composition may be inhomogeneous and anisotropic for the transport quantities, and the properties of the materials are arbitrary functions of temperature field. Since the maximum thermal efficiency of the device is independent of the shape of the leg of a thermoelectric element [33], in this study its shape and size are immaterial. The geometry and physical property of the n-leg generally differ from those of p-leg. Here, we cast the problem using control volume approach along with the method of average parameters [29].

5.3 Control Volume Formulation of a Single Thermoelectric Element

We seek the temperature distribution along the device leg, as it is one of the chief importances for the evaluation of thermal efficiency of the device. Under steady-state condition for the divergence of the flux vector, total energy remains constant along any coordinate direction of a dimensional space. With reference to Fig. 5.2, specializing along x-direction for each leg we obtain [34]

$$TJ_x\left(\frac{\partial\alpha}{\partial x}\right)_T + \tau J_x\frac{\mathrm{d}T}{\mathrm{d}x} - \rho J_x^2 - \frac{\mathrm{d}}{\mathrm{d}x}\left(\kappa\frac{\mathrm{d}T}{\mathrm{d}x}\right) = 0 \tag{5.1}$$

where J_x is the electrical current density vector along x-direction, T is the temperature distribution function, κ is the thermal conductivity of the conductor, ρ is the electrical resistivity of the conductor α, and τ are the Seebeck and the Thomson coefficients, respectively.

The solution of this equation for temperature distribution demands a specification of the dependence of α, κ, ρ, and τ on x or T. One viable approximation consists of replacing all transport coefficients by their averages [29]. In this spirit, the first term in Eq. (5.1) drops out and we arrive at the equation

$$\langle\kappa\rangle\frac{\mathrm{d}^2T}{\mathrm{d}x^2} - \langle\tau\rangle J_x\frac{\mathrm{d}T}{\mathrm{d}x} + \langle\rho\rangle J_x^2 = 0 \tag{5.2}$$

where the symbol $\langle\rangle$ represents an averaged quantity.

Before attempting to solve the resulting simplified equation, it is to be noted that the approximation method is valid only if

$$T_{HC} \approx T_{LC} \tag{5.3a}$$

but

$$T_{HC} > T_{LC} \tag{5.3b}$$

such that the temperature difference across the thermoelectric element $\Delta T = T_{HC} - T_{LC} > 0$. These mathematical restrictions are of little practical interest, since for operation of the device at higher efficiency, temperature difference should be as high as possible on the whole of the thermoelectric device. On the contrary, for very high temperature, the phenomenological representation of irreversible process is inappropriate. Hence, the assumption of negligible temperature gap is consistent with the physical theory developed in the literature [35]. In the real world of engineering design, it represents a cascaded system where power generation takes place discretely in successive stages in series with each other and the power is extracted at each stage. With the increase in the number of modules, the temperature gap across any individual module is reduced and the discrete power generation mimics the continuous power production from a single module.

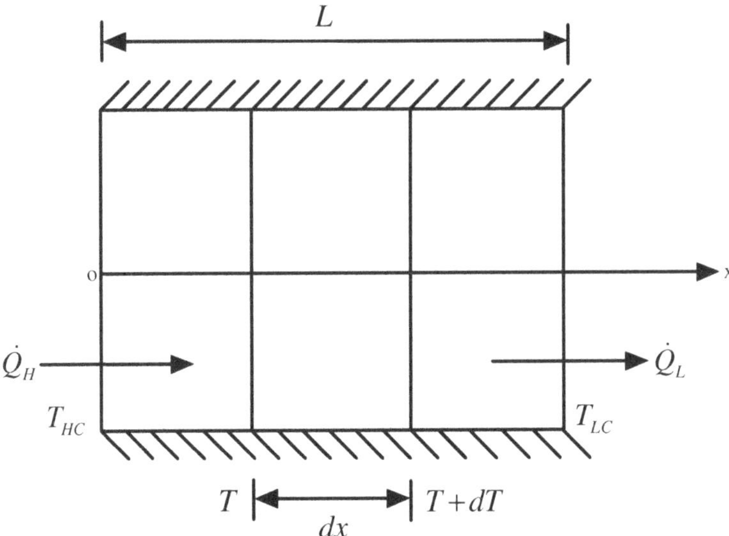

Fig. 5.2 A cascading thermoelectric element exposed to simultaneous heat and current flow

Now, we nondimensionalize Eq. (5.2) using

$$\theta = \frac{T - T_{LC}}{T_{HC} - T_{LC}} = \frac{T - T_{LC}}{\Delta T} \tag{5.4a}$$

and

$$\xi = \frac{x}{L}. \tag{5.4b}$$

The resulting equation takes the form

$$\frac{d^2\theta}{d\xi^2} - \Lambda \frac{d\theta}{d\xi} + \lambda = 0 \tag{5.5}$$

where

$$\Lambda = \frac{\langle \tau \rangle J_x L}{\langle \kappa \rangle} \tag{5.6a}$$

and

$$\lambda = \frac{\langle \rho \rangle (J_x L)^2}{\langle \kappa \rangle \Delta T}. \tag{5.6b}$$

The boundary conditions transform into

$$\theta = 1 \text{ at } \xi = 0 \tag{5.7a}$$

and

$$\theta = 0 \text{ at } \xi = 1. \tag{5.7b}$$

Solution of Eq. (5.5) subjected to boundary conditions (5.7a) and (5.7b) reads as

$$\theta^* = \frac{\lambda}{\Lambda}\xi + \left[\frac{1+\frac{\lambda}{\Lambda}}{1-\exp(\Lambda)}\right]\exp(\Lambda\xi) + \frac{\exp(\Lambda\xi)+\frac{\lambda}{\Lambda}}{\exp(\Lambda)-1}. \tag{5.8}$$

Now, we would like to locate the regime of maximum temperature. This is an important observation when we mimic a thermoelectric device with that of heat engine [11]. In a finite-time heat engine model, there is a continuous variation of temperature from heat source to heat sink along the physical path of energy transport. When both the legs of the thermoelectric device are of the same length, the location of maximum temperature in either of the leg of the thermoelectric generator is obtained by setting $\frac{d\theta^*}{d\xi} = 0$ which yields

$$\xi^* = \frac{1}{\Lambda}\ln\left\{\frac{1}{1+\frac{\Lambda}{\lambda}}\left[\frac{\exp(\Lambda)-1}{\Lambda}\right]\right\}. \tag{5.9}$$

Next, we would like to prescribe some design conditions, which will cause the temperature maximum to pass through the geometrical midpoint of the module of a cascaded thermoelectric device. Each individual module can be thought of as an independent heat engine or one-dimensional insulation system. For a narrow temperature gap across the module, the temperature maximum passes through the midpoint of the device and experiences a minimum entropy generation or equivalently maximum efficiency condition [36]. From definition (5.6a) the imposition of the design criterion $\Lambda \to 0$ leads to the specification of maximum permissible length of any individual thermoelectric module as

$$L_{\text{max}} = \frac{\langle\kappa\rangle}{\langle\tau\rangle}\cdot\frac{1}{J_x}. \tag{5.10a}$$

Again, the design prescription $\frac{\Lambda}{\lambda} \to 0$ stipulates from definitions (5.6a) and (5.6b) the minimum permissible length of the individual thermoelectric module as

$$L_{\text{min}} = \frac{\langle\tau\rangle}{\langle\rho\rangle}\cdot\frac{\Delta T}{J_x}. \tag{5.10b}$$

Thus, it is to be observed that the minimum length of the device arm depends on the applied temperature gap, whereas maximum permissible length is devoid of dependence of such imposed temperature gradient.

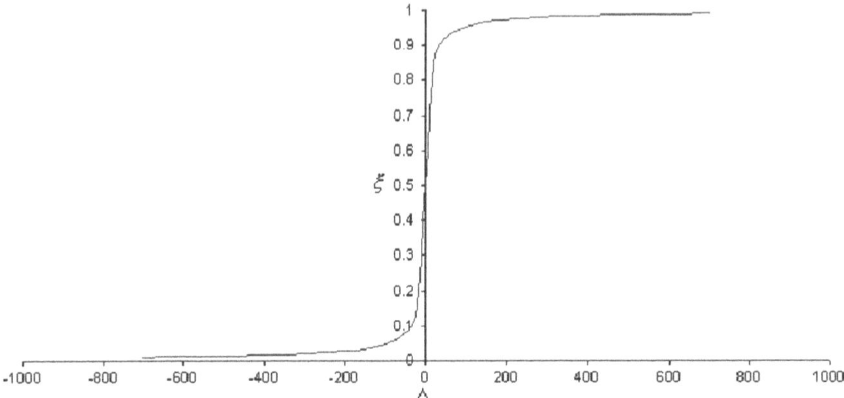

Fig. 5.3 Location of maximum temperature in concurrence with Thomson heat

Equation (5.10b) as a design criterion transforms Eq. (5.9) into the form

$$\xi^* = \frac{1}{\Lambda}\ln\left[\frac{\exp(\Lambda) - 1}{\Lambda}\right]. \tag{5.11a}$$

Equation (5.11a) is plotted in Fig. 5.3, which shows that for $\Lambda \to 0$, the relation between ξ and Λ is linear. Expanding the left side of Eq. (5.11a) analytically around the singular point $\Lambda = 0$ and then passing to the limit, we have

$$\lim_{\Lambda\to 0}\xi^* = \lim_{\Lambda\to 0}\frac{1}{\Lambda}\ln\left\{\frac{1}{\Lambda}\left[\left(1 + \Lambda + \frac{\Lambda^2}{2!} + \frac{\Lambda^3}{3!} + \cdots\cdots\cdots\right) - 1\right]\right\} = \frac{1}{2}. \tag{5.11b}$$

Thus, Eq. (5.11b) clearly demonstrates that for $\Lambda = 0$, temperature maximum passes through the geometric midpoint of the conductor as the electrical current changes the direction. As long as Eqs. (5.3a) and (5.3b) are valid, the result obtained in Eq. (5.11b) is physically realistic.

In order to construct a cascaded system, the length of the first junction should be at half the total permissible length of the assembly of the thermocouples. The geometric midpoint will act as a heat source for the next junction and so on. Figure 5.4 schematically represents an assembly of cascaded thermoelectric generators.

The construction has two limiting conditions on the length of each module stipulated by Eqs. (5.10a) and (5.10b). When on account of thermophysical properties, imposed temperature gradient, permissible current, and space restriction, the maximum and minimum lengths of the installation are known, the architecture of the assembly can be proceeded either from small to large or vice versa. In Fig. 5.4 L_1L_1' represents the highest permissible length of a thermoelectric module. The second, third, and fourth constructs of the assembly are L_2L_2', L_3L_3', and L_4L_4', respectively. The resulting structure looks like a *T*-shaped region in space. Thus, unlike fractals [37] and biomimetics [38], the optimum structure provides the physics of the organization.

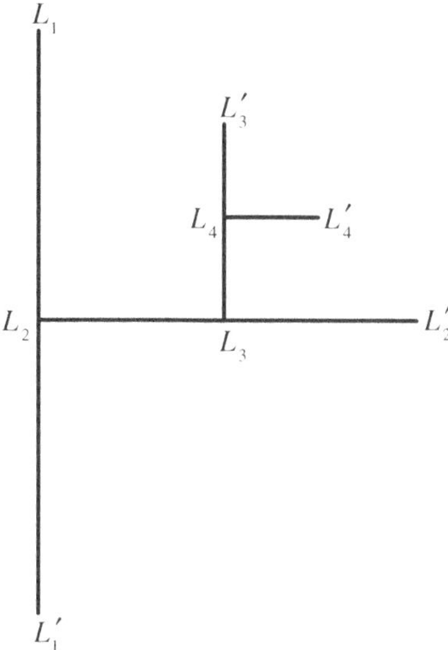

Fig. 5.4 Fractal-like but deterministic (constructal) assembly of cascaded thermoelectric generator modules

Heat flow toward the hot end is calculated by invoking Fourier law of heat conduction as

$$\dot{Q}_H^* = -\kappa A \frac{\partial T}{\partial x}\bigg|_{x=0} = -\dot{Q}_k \frac{\partial \theta^*}{\partial \xi}\bigg|_{\xi=0} = \dot{Q}_J \left[\frac{\Lambda - \exp(\Lambda) + 1}{\Lambda \exp(\Lambda) - \Lambda}\right] \tag{5.12}$$

where the conducted heat through cross-sectional area A and of length L is

$$\dot{Q}_k = \frac{\kappa A \Delta T}{L} \tag{5.13}$$

and the Joulean heat source of cross-sectional area A and length L is

$$\dot{Q}_J = \rho J_x^2 A L \tag{5.14}$$

such that as $\Lambda \to 0$

$$\lambda \approx \frac{\dot{Q}_J}{\dot{Q}_k}. \tag{5.15}$$

We calculate the ratio $\left|\frac{\dot{Q}_H^*}{\dot{Q}_J}\right|$ in the limit $\Lambda \to 0$ to examine what proportion of Joulean heat moves to the hot end. On calculating the limit, using L' Hospital's theorem, we have

$$\lim_{\Lambda \to 0}\left|\frac{\dot{Q}_H^*}{\dot{Q}_J}\right| = \lim_{\Lambda \to 0}\left|\frac{\Lambda - \exp(\Lambda) + 1}{\Lambda \exp(\Lambda) - \Lambda}\right| = \frac{1}{2}. \tag{5.16}$$

From Eq. (5.16) we observe that exactly half the Joulean heat proceeds to the hot end. The first law of thermodynamics asserts that precisely 50 % of the Joulean heat contributes to the cold junction.

If the electric current changes its direction, the Thomson heat also changes its sign. This implies that if two parallel conductors of almost the same geometrical and thermoelectrical attribute are placed in communication with a single reservoir and if the same strength of current flows in opposite directions through the two conductors, the Thomson heat so generated by one conductor agrees nearly with the Thomson heat absorbed by the other. Thus, the effect of Thomson heat is almost nullified. Hence, the net heat transfer interaction with the thermal reservoir comprises rejecting only two Joulean heating rates generated by the two conductors. From Eq. (5.6a), it can be noted that the Thomson effect need not be absent even for a vanishingly small value of the parameter Λ, as for any individual cascading member the passing current and length of the element are small and heat transfer irreversibility phenomenon overwhelms Thomson effect. Another mathematical way of looking at the problem is the imposition of the restrictions that κ, ρ, and α are constants. Then both $\left(\frac{\partial \alpha}{\partial x}\right)_T$ and $\tau = T\left(\frac{\partial \alpha}{\partial T}\right)_x$ vanish such that Eq. (5.2) further reduces to

$$\kappa_0 \frac{d^2 T}{dx^2} + \rho_0 J_x^2 = 0 \tag{5.17}$$

where the constant values are indicated by the subscript zero.

At first sight, it seems that the above argument waives the imposition of the restriction that the temperature difference should be small if we ignore the origin of Eq. (5.17). It is to be noted that the constancy of these thermophysical and electrical properties demand in turn the narrow temperature range of operation of the individual element of the device. Alternatively, Eq. (5.17) can be readily obtained by applying the first law of thermodynamics for a control volume where conduction of heat takes place with distributed heat source without introducing the formalism of irreversible thermodynamics. Unlike thermionic device, a cascaded thermoelectric element works under narrow temperature range and hence we can neglect the very effect of radiation and convection. The distribution of temperature along a thin conductor under the influence of high current involving radiative and conductive transfer is reported in the literature [39–42] from a different perspective. Contrary to Thomson heat consideration above, in this limiting case we have the liberty to formulate the boundary conditions as follows:

$$T = T_{HC} \text{ at } x = 0, \tag{5.18a}$$

$$T = T_{LC} \text{ at } x = L, \tag{5.18b}$$

and

$$T_{HC} > T_{LC}. \tag{5.18c}$$

The absolute value of both T_{HC} and T_{LC} is but small.

Nondimensional solution of Eq. (5.17) subjected to the boundary conditions (5.18a), (5.18b), and (5.18c) is

$$\theta_* = (1 - \xi) + \frac{\zeta}{2}(\xi - \xi^2) \tag{5.19}$$

where

$$\zeta = \frac{\rho_0 (J_x L)^2}{\kappa_0 \Delta T} = \frac{\dot{Q}_J}{\dot{Q}_k}. \tag{5.20}$$

Once again, the location of maximum temperature is obtained by setting $\frac{\mathrm{d}\theta_*}{\mathrm{d}\xi} = 0$ of Eq. (5.19) and the final result is

$$\xi_* = \frac{1}{2}\left(1 - \frac{2}{\zeta}\right). \tag{5.21}$$

Passing to the limit $\zeta \to \infty$ in Eq. (5.21)

$$\lim_{\zeta \to \infty} \xi_* = \lim_{\zeta \to \infty} \frac{1}{2}\left(1 - \frac{2}{\zeta}\right) = \frac{1}{2} \tag{5.22}$$

we notice that ξ_* asymptotically approaches the finite value $\frac{1}{2}$. Hence, the numerical value of ζ should be high for temperature maximum to occur at the geometrical middle of the conductor. Thus, even for the idealized situation when the thermoelectric element behaves like a resistor under the influence of low current, the placement of the second junction begins at the middle as if the thermoelectric element were not cascaded.

Heat flow inward the high temperature side is given by

$$\dot{Q}_{*H} = -\kappa A \frac{\partial T}{\partial x}\bigg|_{x=0} = -\dot{Q}_k \frac{\partial \theta_*}{\partial \xi}\bigg|_{\xi=0} - \dot{Q}_J\left[\frac{1}{\zeta} - \frac{1}{2}\right]. \tag{5.23}$$

We evaluate the ratio $\left|\frac{\dot{Q}_{*H}}{\dot{Q}_J}\right|$ in the limit $\zeta \to \infty$ as

$$\lim_{\zeta \to \infty}\left|\frac{\dot{Q}_{*H}}{\dot{Q}_J}\right| = \lim_{\zeta \to \infty}\left|\frac{1}{\zeta} - \frac{1}{2}\right| = \frac{1}{2}. \tag{5.24}$$

Equation (5.24) confirms that only half the Joulean heat goes to the hot end. Energy balance states that sharply half the Joulean heat arrives at the cold end.

5.4 Control Volume Formulation for the Complete Thermoelectric Device

In order to maintain consistency with the standard notation of analysis prevailing in the literature, we define the relationship between electrical resistance and resistivity, thermal conductance, and conductivity of the thermoelectric element introduced in the above section. Electrical resistance R is related to its counterpart resistivity ρ through

$$R = \frac{\rho L}{A}. \tag{5.25a}$$

Thermal conductance K is dependent on conductivity κ as

$$K = \frac{\kappa A}{L}. \tag{5.25b}$$

For control volume formulation of the integrated thermoelectric device as shown in Fig. 5.5, we employ Newton's law of cooling [43]. The first law of thermodynamics analysis neglecting Thomson effect enables us to write down the following heat transport equations in algebraic forms [11]. Finite-time heat transfer rate to the hot junction $\dot{Q}_H$ is given by

$$\dot{Q}_H = K_H(T_H - T_{HC}) = \alpha I T_{HC} + K(T_{HC} - T_{LC}) - F_H I^2 R \tag{5.26}$$

where K and K_H are the thermal conductances across the reversible compartment and the hot junction, respectively. T_H is the temperature of the high temperature source and T_{HC} is that of thermoelectric element such that $T_{HC} \leq T_H$. Fraction of Joulean heat entering into the hot junction is F_H. Equation (5.26) can be rearranged as

$$(K + K_H + \alpha I)T_{HC} - KT_{LC} - \left(K_H T_H + F_H I^2 R\right) = 0. \tag{5.26a}$$

Similarly, finite-time heat transfer rate to the cold junction $\dot{Q}_L$ is obtained as

$$\dot{Q}_L = K_L(T_{LC} - T_L) = \alpha I T_{LC} + K(T_{HC} - T_{LC}) + F_L I^2 R \tag{5.27}$$

where K_L is the thermal conductance across the cold junction. T_L is the temperature of the low temperature sink and T_{LC} is that of thermoelectric component such that $T_{LC} \geq T_L$. Fraction of Joulean heat entering into the cold junction is F_L. Equation (5.27) can be rewritten as

$$KT_{HC} - (K + K_L - \alpha I)T_{LC} + \left(K_L T_L + F_L I^2 R\right) = 0. \tag{5.27a}$$

For the Joulean heat distribution it is obvious that

$$F_H + F_L = 1. \tag{5.28}$$

Now, the system of Eqs. (5.26a), (5.27a), and (5.28) has four variables T_{HC}, T_{LC}, F_L, and F_H rendering single degree of freedom. Choosing F_H to be that degree of freedom, we solve for T_{HC} and T_{LC} to obtain

$$T_{HC} = \frac{K[(K_H T_H + K_L T_L) + I^2 R] + (K_L - \alpha I)(K_H T_H + F_H I^2 R)}{K(K_H + K_L) + (K_H + \alpha I)(K_L - \alpha I)} \tag{5.29a}$$

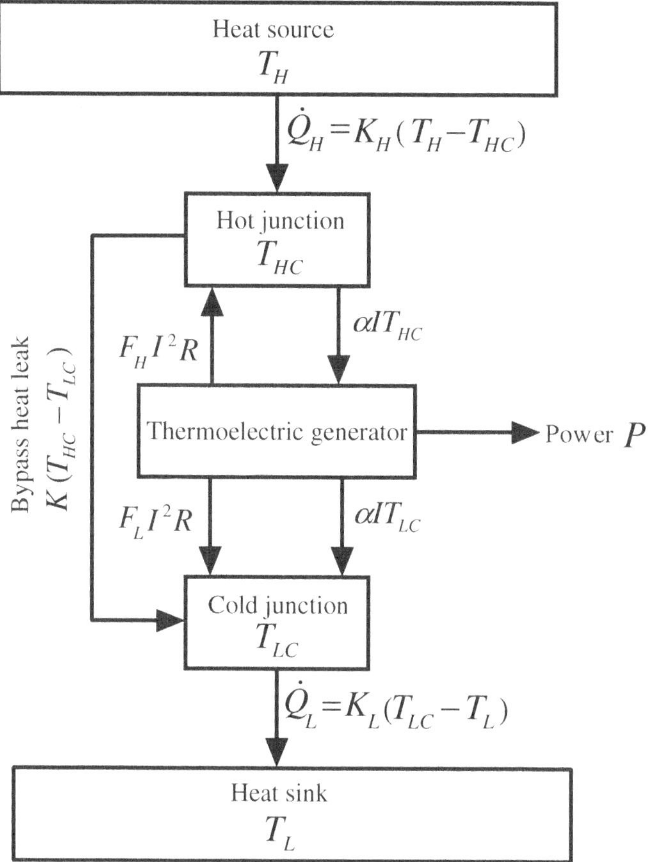

Fig. 5.5 Schematic diagram of a cascading finite-time thermoelectric power generator comprising two differentially heated thermoelectric elements

and

$$T_{LC} = \frac{K[(K_H T_H + K_L T_L) + I^2 R] + (K_H + \alpha I)(K_H T_H + F_H I^2 R)}{K(K_H + K_L) + (K_H + \alpha I)(K_L - \alpha I)}. \tag{5.29b}$$

Next, we proceed to seek a possible set of solutions for the assumed unknown variable F_H or F_L. Among many other methodologies [44], we devise our own solution strategy based on the symmetry of the problem as follows. Eliminating T_{HC} between Eqs. (5.26a) and (5.27a) and providing an expression for T_{LC} from Eq. (5.29b), we obtain

$$\begin{aligned}&[K(K_H T_H + K_L T_L) + K_H K_L T_L] + \alpha K_L T_L I + R[K + F_L K_H] I^2 + \alpha F_L R I^3 \\ &= [K(K_H T_H + K_L T_L) + K_H K_L T_L] + \alpha K_L T_L I + R[F_L(K + K_H) + F_H K] I^2 + \alpha F_L R I^3.\end{aligned} \tag{5.30}$$

Comparing like powers of I, we have for the term I^2

$$K + F_L K_H = (K + K_H)F_L + KF_H. \tag{5.31}$$

Analogously, T_{LC} eliminant of Eqs. (5.26a) and (5.27a) with the insertion of the expression for T_{HC} from Eq. (5.29a), we get

$$\begin{aligned}&[K(K_H T_H + K_L T_L) + K_L K_H T_H] - \alpha K_H T_H I + R[K + F_H K_L]I^2 - \alpha F_H R I^3 \\ &= [(K + K_L)K_H T_H + KK_L T_L] - \alpha K_H T_H I + R[F_H(K + K_L) + F_L K]I^2 - \alpha F_H R I^3.\end{aligned} \tag{5.32}$$

Equating similar powers of I on both sides, we have for I^2

$$K + F_H K_L = (K + K_L)F_H + KF_L. \tag{5.33}$$

Any particular solution of two identities (5.31) and (5.33) must have general functional form involving K, K_H, and K_L, i.e.,

$$F_H = F_H(K, K_H, K_L) \tag{5.34a}$$

and

$$F_L = F_L(K, K_H, K_L). \tag{5.34b}$$

Following symmetry, we assume a trial solution of the form

$$F_H = \frac{1}{2}\left(\frac{K_H K_L + KK_H + KK_H}{K_H K_L + KK_H + KK_L}\right) = \frac{1}{2}\left(\frac{K_H K_L + 2KK_H}{K_H K_L + KK_H + KK_L}\right). \tag{5.35a}$$

Employing Eq. (5.35a) in Eq. (5.28), we obtain

$$F_L = \frac{1}{2}\left(\frac{K_H K_L + KK_L + KK_L}{K_H K_L + KK_H + KK_L}\right) = \frac{1}{2}\left(\frac{K_H K_L + 2KK_L}{K_H K_L + KK_H + KK_L}\right). \tag{5.35b}$$

Substituting Eqs. (5.35a) and (5.35b) into the identity (5.31), we have for both sides a common expression

$$C_1 = \frac{4KK_H K_L + 2K^2(K_H + K_L) + K_H^2 K_L}{2(K_H K_L + KK_H + KK_L)}. \tag{5.36a}$$

Similarly, inserting Eqs. (5.35a) and (5.35b) into the other identity (5.33), we obtain another common expression

$$C_2 = \frac{4KK_H K_L + 2K^2(K_H + K_L) + K_H K_L^2}{2(K_H K_L + KK_H + KK_L)}. \tag{5.36b}$$

Equations (5.36a) and (5.36b) confirm that Eqs. (5.35a) and (5.35b) are a set of possible solutions for the identities (5.31) and (5.33). Further, by simple inspection we observe that Eqs. (5.31) and (5.33) admit the following numerical values:

$$F_H = 0 \text{ and } F_L = 1, \tag{5.37a}$$

$$F_H = 1 \text{ and } F_L = 0, \tag{5.37b}$$

and

$$F_H = \frac{1}{2} \text{ and } F_L = \frac{1}{2}. \tag{5.37c}$$

Now, we will examine for what combinations of K, K_H, and K_L these numeric values are returned for the functional relations (5.31) or (5.33). Equations (5.35a) and (5.35b) along with Eq. (5.37a) say that

$$K_H K_L - \quad 2KK_H. \tag{5.38a}$$

For Eq. (5.37b) to be tantamount with Eqs. (5.35a) and (5.35b), one requires that

$$K_H K_L = -2KK_L. \tag{5.38b}$$

Equivalency of Eqs. (5.35a) and (5.35b) with Eq. (5.37c) demands that

$$K_H = K_L. \tag{5.38c}$$

Since, K, K_H, and K_L are all nonnegative quantities, only Eqs. (5.37c) and (5.38c) are physically realistic. Thus, the necessary and sufficient condition for equipartition of Joulean heat produced is the equipartition of conductance allocations between high temperature heat source and low temperature heat sink. Furthermore, when one of the three conductances runs to a very high value leaving the other two in a moderate range, Joulean heat distribution again becomes unequal.

5.5 Consequences of Equipartitioned Joulean Heat

The better the thermoelectric material for direct energy conversion, the higher should be the value of the dimensionless group zT where T is the average absolute temperature and z is the figure-of-merit of the thermoelectric material. This dimensionless parameter for a semimetal or semiconductor can be expressed, in general, as [45]

$$zT = \alpha^2 T\left(\frac{\sigma}{\kappa}\right) \tag{5.39a}$$

where σ is the electrical conductivity and the reciprocal of electrical resistivity. Thermal conductivity can further be treated as the cumulative effect of electrical conductivity κ_e and lattice thermal conductivity κ_l such that

$$\kappa = \kappa_e + \kappa_l. \tag{5.39b}$$

For higher orders of magnitude of the dimensionless group on the left side of Eq. (5.39a) in conjunction with Eqs. (5.6a) and (5.6b) and with the conditions $\Lambda \to 0$ and $\zeta \to \infty$, we may stipulate that

$$\tau \to 0+,\ \kappa \to 0+,\ \rho \to 0+,\ \text{and}\ \Delta T \to 0+. \tag{5.40}$$

Thus, we conclude that the present analysis is valid even for the finite temperature difference and the effect of Thomson heat can be neglected for a good quality thermoelectric material.

Equations (5.11b) and (5.16) together exhibit an interesting thermodynamic property of a thermoelectric element. For temperature maximum to occur at the geometric center of a thermoelectric element, exactly half the Joulean heat approaches to the hot end and the other half to the cold side. Similar observation is repeated through Eqs. (5.22) and (5.24).

From Eqs. (5.37c) and (5.38c), we learn that both the hot and cold junctions experience half the net Joulean effect produced which demands in turn equal conductance allocation on both sides. Equation (5.38c) also prompts the fact that both K_H and K_L are finite and hence the following proposition holds:

$$K_H + K_L = C \tag{5.41a}$$

where C is some finite constant. In the engineering literature, conductance is denoted as a product of overall heat transfer coefficient U and related surface area A. Thus, Eq. (5.41a) can be rewritten as

$$U_H A_H + U_L A_L = C \tag{5.41b}$$

where the subscripts H and L refer to the high temperature source and low temperature sink, respectively. Allocation of heat exchanger inventory was extensively investigated in connection with the optimization of refrigeration and power production both from the thermodynamic [46–48] and thermoeconomic [49] viewpoints. But the final result of an optimization problem depends on the nature of imposed constraint [30]. Klein [48] considered the constraint of the type

$$\varepsilon_H U_H + \varepsilon_L U_L = C_K \tag{5.41c}$$

where ε is the effectiveness of the heat exchanging equipment and C_K is a constant. Based on the notion that both conductance and entropy generation have the same dimension, Ait-Ali [46] conceived a condition of the form

$$\frac{\dot{Q}_H}{T_H - T_{HC}} + \frac{\dot{Q}_L}{T_{LC} - T_L} = C_A \tag{5.41d}$$

where C_A is a parametric constant. On the basis of the total cost conservation of heat exchanger installation, Antar and Zubair [49] framed a relation as

$$\gamma_H U_H A_H + \gamma_L U_L A_L = C_Z \tag{5.41e}$$

where γ is the unit conductance cost and C_Z has a fixed value. Bejan [47] considered the maximization of power production of heat engines and refrigeration load in refrigerators with two heat reservoirs considering the total area constraint for the heat exchangers on following the equation

$$A_H + A_L = C_B \tag{5.41f}$$

where C_B is a constant due to some resource constraint. Treating Eq. (5.41b) as well as Eq. (5.41f), Bejan concluded that in both the cases either quantitative or qualitative equipartition of thermal conductance inventory is valid. Similar results have been echoed in other works [47–49]. Thus, we propose that either Eq. (5.41a) or Eq. (5.41b) is the most natural constraint for such category of optimization problems. It is to be noted that Eq. (5.41a) or Eq. (5.41b) is but the law of motive force.

5.6 Discussions

Creditably, the debatable concept of endoreversibility [50–52] in finite-time thermodynamics can be mitigated by incorporating some irreversibility factors to the reversible compartment sandwiched between two irreversible chambers. Consideration of bypass heat leak is a compensating measure to this direction. Joulean heating present in a thermoelectric generator itself is an inherent source of irreversibility and tantamount to the frictional loss in a heat engine. The discrimination between frictional heat leak and heat loss due to finite rate heat transfer was first put forward in a work by Andresen et al. [53]. In this chapter, bypass heat leak is identified as a major contribution to the measure of irreversibility. The sufficiency of bypass heat leak consideration in engine modeling is an established practice [11].

Our analysis shows that in a thermoelectric generator, Thomson effect may be neglected in one limiting case or may not be negligible in another limiting situation even for a vanishingly small value of a certain nondimensional parameter Λ. A very high value of another dimensionless factor ζ recognizes a better figure-of-merit and the operation of the thermoelectric device as cascaded system over a small but finite temperature gap.

Three parameters Λ, λ, and ζ so identified are responsible for temperature maximum to pass through the geometrical midpoint of the one-dimensional physical device. This observation is in conformity with the principle of insulation design and the broader sense of engine modeling. The parallelism between the design principle of heat engine, heat exchanger, and refrigerator to that of insulation system was established by a pioneering work of Bejan [36]. For the most efficient system a stack of insulation is cooled midway. Similarly, the calculation of mid-point temperature is of intrinsic importance also in the application of thermionic elements where an interesting phenomenon occurs in the middle of the conductor: the temperature reaches extremum, remains constant there, and Joulean heating and radiative heat transfer takes an equal share of the feeded energy [39–42].

Imposing the design prescription $\Lambda \to 0$ leads to the limit of maximum permissible length of any individual module of the cascaded thermocouple. Another design criterion $\frac{\Lambda}{\lambda} \to 0$ stipulates the maximum permissible length of such module. It is to be noticed that the maximum allowable length depends explicitly only on the thermoelectric properties of the material and the value of the passing current, whereas the minimum permissible length additionally depends on applied temperature gradient.

For the ideal values of these three parameters Λ, λ, and ζ, it is interesting to report that exactly half the Joulean heat *flows* into the hot end and half to the cold junction. It is to be noted that this is not equivalent to state that half of the Joulean heat affects either end [13].

Control volume formulation for the integrated device disseminates that exactly half the Joulean heat *affects* both the junctions. This observation is duly supported by experimental evidence [13, 14, 54] in a similar class of thermoelectric devices.

It is conditioned that for finite bypass heat leak, optimal conductance allocation is equipartitioned between high and low temperature sides when Joulean heat *affects* both the junctions equally. This instance of equipartition also conforms with the corollary [55] of constructal theory as well as the law of motive force.

The physical solution presented in this work actually pertains to Steiner-like [56, 57] problems in mathematics that has defeated the fastest computers [58]. The elegant variational solution [59] of such branching network is practically very complex even in one dimension [60].

References

1. Bejan, A.: Advanced Engineering Thermodynamics, pp. 665–682. Wiley, New York (2006)
2. Bridgman, P.W.: Thermoelectric phenomena in crystals and general electrical concepts. Phys. Rev. **31**, 221–235 (1928)
3. Goldsmid, H.J.: Thermoelectric Refrigeration. Plenum, New York (1964)
4. Heikes, R.R., Ure Jr., R.W. (eds.), Mullin, A.A. (rev.).: Thermoelectricity: science and engineering. Am. J. Phys. **30**, 78 (1962)
5. Ioffe, A.F.: The revival of thermoelectricity. Sci. Am. **199**, 31–37 (1958)
6. Ioffe, A.F.: Semiconductor Thermoelements and Thermoelectric Cooling. Infosearch Limited, London (1957)
7. Thomson, W.: Thermoelectric currents. In: Mathematical and Physical Papers-I, pp. 232–291. Cambridge University Press, Cambridge (1882)

8. Wiśniewski, S., Staniszewski, B., Szymanik, R.: Thermodynamics of Nonequilibrium Processes (trans: Lepa, E.), pp. 128–180. D. Reidel, Boston (1976)
9. Gupta, V.K., Gauri, S., Sarat, B., Sharma, N.K.: Experiment to verify the second law of thermodynamics using a thermoelectric device. Am. J. Phys. **52**, 625–628 (1984)
10. Yan, Z., Chen, J.: Comment on "Generalized power versus efficiency characteristics of heat engines: the thermoelectric generator as an instructive illustration". Am. J. Phys. **61**, 380 (1993)
11. Gordon, J.M.: Generalized power versus efficiency characteristics of heat engines: the thermoelectric generator as an instructive illustration. Am. J. Phys. **59**, 551–555 (1991)
12. Gordon, J.M.: A response to Yan and Chen's "Comment on 'Generalized power versus efficiency characteristics of heat engines: the thermoelectric generator as an instructive illustration'". Am. J. Phys. **61**, 381 (1993)
13. Luke, W.H.: Reply to experiment in thermoelectricity. Am. J. Phys. **28**, 563 (1960)
14. Noon, J.H., O'Brien, B.J.: Sophomore experiment in thermoelectricity. Am. J. Phys. **26**, 373–375 (1958)
15. Andresen, B., Salamon, P., Berry, R.S.: Thermodynamics in finite time. Phys. Today **37**, 62–70 (1984)
16. Chen, J.: The maximum power output and maximum efficiency of an irreversible Carnot heat engine. J. Phys. D Appl. Phys. **27**, 1144–1149 (1994)
17. Curzon, F.L., Ahlborn, B.: Efficiency of a Carnot engine at maximum power output. Am. J. Phys. **43**, 22–24 (1975)
18. De Mey, G., De Vos, A.: On the optimum efficiency of endoreversible thermodynamic processes. J. Phys. D Appl. Phys. **27**, 736–739 (1994)
19. De Vos, A.: Reflections on the power delivered by endoreversible engines. J. Phys. D Appl. Phys. **20**, 232–236 (1987)
20. Gordon, J.M.: Maximum power point characteristics of heat engines as a general thermodynamic problem. Am. J. Phys. **57**, 1136–1142 (1989)
21. Yan, Z., Chen, L.: The fundamental optimal relation and the bounds of power output and efficiency for an endoreversible Carnot engine. J. Phys. A Math. Gen. **28**, 6167–6175 (1995)
22. Månsson, B.Å.: Thermodynamics and economics. In: Sieniutycz, S., Salamon, P. (eds.) Finite-Time Thermodynamics and Thermoeconomics. Taylor & Francis, New York (1991)
23. Rubin, M.H.: Optimal configuration of a class of irreversible heat engines-I. Phys. Rev. A **19**, 1272–1276 (1977)
24. Bejan, A., Paynter, H.M.: Solved Problems in Thermodynamics. Problem VIID, MIT, Cambridge (1976)
25. El-Wakil, M.M.: Nuclear Power Engineering, pp. 162–165. McGraw-Hill, New York (1962)
26. Lu, P.C.: On optimal disposal of waste heat. Energy **5**, 993–998 (1980)
27. Novikov, I.I.: The efficiency of atomic power stations. J. Nucl. Energy II **7**, 125–128 (1958)
28. Bejan, A.: Shape and Structure, from Engineering to Nature. Cambridge University Press, Cambridge (2000)
29. Sherman, B., Heikes, R.R., Ure Jr., R.W.: Calculation of efficiency of thermoelectric devices. J. Appl. Phys. **31**, 1–16 (1960)
30. De Vos, A., Desoete, B.: Equipartition principle in finite-time thermodynamics. J. Non-Equilib. Thermodyn. **25**, 1–13 (2000)
31. Pramanick, A.K., Das, P.K.: Constructal design of a thermoelectric device. Int. J. Heat Mass Transf. **49**, 1420–1429 (2006)
32. Pramanick, A.K.: Equipartition of Joulean heat in thermoelectric generators. In: Rocha, L.A.O., Lorente, S., Bejan, A. (eds.) Constructal Law and the Unifying Principle of Design. Springer, New York (2013)
33. Boerdijk, A.H.: Contribution to a general theory of thermocouples. J. Appl. Phys. **30**, 1080–1083 (1959)
34. Harman, T.C., Honig, J.M.: Thermoelectric and Thermomagnetic Effects and Applications, p. 276. McGraw-Hill, New York (1967)

35. De Groot, S.R.: Thermodynamics of Irreversible Processes, pp. 141–162. Wiley-Interscience, New York (1952)
36. Bejan, A.: Entropy Generation Through Heat and Fluid Flow, pp. 173–187. Wiley, New York (1982)
37. Kadanoff, L.P.: Fractals: where's the physics? Phys. Today **39**, 6–7 (1986)
38. McMahon, T.A., Kronauer, R.E.: Tree structures: deducing the principle of mechanical design. J. Theor. Biol. **59**, 443–466 (1976)
39. Jain, S.C., Krishnan, K.S.: The distribution of temperature along a thin rod electrically heated *in vacuo*. I. Theoretical. Proc. R. Soc. Lond. A **222**, 167–180 (1954)
40. Jain, S.C., Krishnan, K.S.: The distribution of temperature along a thin rod electrically heated *in vacuo*. II. Theoretical. Proc. R. Soc. Lond. A **225**, 1–7 (1954)
41. Jain, S.C., Krishnan, K.S.: The distribution of temperature along a thin rod electrically heated *in vacuo*. III. Experimental. Proc. R. Soc. Lond. A **225**, 7–18 (1954)
42. Jain, S.C., Krishnan, K.S.: The distribution of temperature along a thin rod electrically heated *in vacuo*. IV. Many useful formulae verified. Proc. R. Soc. Lond. A **225**, 19–32 (1954)
43. Salamon, P., Nitzan, A.: Finite time optimizations of a Newton's law Carnot cycle. J. Chem. Phys. **74**, 3546–3560 (1981)
44. Rektorys, K. (ed.): Survey of Applicable Mathematics, pp. 70–75. Liffe Books, London (1969)
45. Min, G., Rowe, D.M.: Thermoelectric figure-of-merit barrier at minimum lattice thermal conductivity? Appl. Phys. Lett. **77**, 860–862 (2000)
46. Ait-Ali, M.: Maximum power and thermal efficiency of an irreversible power cycle. J. Appl. Phys. **78**, 4313–4318 (1995)
47. Bejan, A.: Theory of heat transfer-irreversible power plants—II. The optimal allocation of heat exchange equipment. Int. J. Heat Mass Transf. **38**, 433–444 (1995)
48. Klein, S.A.: Design considerations for refrigeration cycles. Int. J. Refrg. **15**, 181–185 (1992)
49. Antar, M.A., Zubair, S.M.: Thermoeconomic considerations in the optimum allocation of heat exchanger inventory for a power plant. Energ. Convers. Manage. **42**, 1169–1179 (2001)
50. Andresen, B.: Comment on "A fallacious argument in the finite time thermodynamic concept of endoreversibility". J. Appl. Phys. **90**, 6557–6559 (2001)
51. Sekulic, D.P.: A fallacious argument in the finite time thermodynamics concept of endoreversibility. J. Appl. Phys. **83**, 4561–4565 (1998)
52. Sekulic, D.P.: Response to "Comment on 'A fallacious argument in the finite time thermodynamics concept of endoreversibility'". J. Appl. Phys. **90**, 6560–6561 (2001)
53. Andresen, B., Salamon, P., Berry, R.S.: Thermodynamics in finite time: extremals for imperfect heat engines. J. Chem. Phys. **66**, 1571–1577 (1977)
54. Logan, J.K., Clement, J.R., Jeffers, H.R.: Resistance minimum of magnesium: heat capacity between 3°K and 13°K. Phys. Rev. **105**, 1435–1437 (1957)
55. Pramanick, A.K., Das, P.K.: Note on constructal theory of organization in nature. Int. J. Heat Mass Transf. **48**, 1974–1981 (2005)
56. Gray, A.: Tubes. Birkhäuser, Boston (2004)
57. Hwang, F.K., Richards, D.S., Winter, P.: The Steiner Tree Problem. Elsevier, London (1992)
58. Bern, M.W., Graham, R.L.: The shortest network problem. Sci. Am. **260**, 84–89 (1989)
59. Rubinstein, J.H., Thomas, D.A.: A variational approach to the Steiner network problem. Ann. Oper. Res. **33**, 481–499 (1991)
60. Ivanov, A.O., Tuzhilin, A.A.: Branching Solutions to One-Dimensional Variational Problems. World Scientific, Philadelphia (2001)

Chapter 6
Real Heat Engine

Since the turn of the century anyone who has set pen to paper in an attempt to advance thermodynamics has come under attack from one quarter or another, and the only thing upon which we all agree that Gibbs was a very smart fellow. So, not knowing what to make of the battles raging around us, we opt for neutrality; we confine our teaching to the substance and style of 19th century thermodynamics. Although this course of action has served us reasonably well and, incidentally, lends to the subject an undeniable charm, at some point we must ask if such a state of affairs is to prevail forever.

M. Feinberg

In this chapter, we turn our attention to the features of a more realistic heat engine, unlike thermoelectric generator, which is considered to be the natural heat engine. In the first place, we abandon the linear heat transfer law for the external heat transfer resistance while adopting a generalized power law. Such a power law is immediately inclusive of linear model representing conventional Newton's law of cooling, phenomenological heat transfer law, and radiative heat transmission mode among a host. The complex index of power law heat transfer duly takes into account the relaxation phenomenon in heat transfer. Bypass heat leak being staged through the mechanical supports of the engine is left to remain as linear. The work producing compartment is no longer considered to be endoreversible, rather it is aptly labeled as irreversible. Realistically, leaving behind the temperatures of the heat source and sink to be the real quantities, the temperatures of the working fluid at the hot end and the cold side are contemplated as complex quantities to take into account the oscillating nature of heat transfer by the fluid flow. As the engineering choice is more restricted to the selection of working fluid, the optimization objective truly remains to be the optimal allocation of heat exchanger inventory alone. The above features of a more realistic engine are still amenable to a closed-form analytical solution through complex analysis by the employment of the law of motive force.

6.1 The Problem

The fundamental contribution of simple models is to provide an estimate of different important parameters of a functional device and to establish a way for the more applied work that will follow in due course of industrial research and

A. K. Pramanick, *The Nature of Motive Force*, Heat and Mass Transfer,
DOI: 10.1007/978-3-642-54471-2_6,

development [1]. In this chapter, many simplistic assumptions commonly adopted for a power plant or heat engine are abandoned making it more actual for the realistic performance. In the following paragraphs we discuss the issues that corner around the design of an actual power plant.

To start with, there are many practical engineering concerns to contemplate in connection with the four-process model optimized by Curzon and Ahlborn [2]. Sadi Carnot's original essay [3] as interpreted graphically and analytically by Émile Clapeyron [4] is a description of a gas contained in a cylinder and piston mechanism that undergoes a cycle of four processes: two quasistatic and isothermal processes interspersed with two quasistatic and adiabatic processes. Recently, Landsberg et al. [5] generalized this cycle, which is characterized by two adiabatics and two heat transfer paths with constant heat capacities. Curzon and Ahlborn added finite thermal resistances between the cylinder and the respective temperature reservoirs and in this way described and optimized a more realistic time-dependent evolution of the cycle. The four-process model of Curzon and Ahlborn and its steady-state counterpart introduced by Novikov [6] and independently by Bejan et al. [7], Andresen et al. [8], and Lu [9] was perused along several lines. These were reviewed by Andresen et al. [10], Wu et al. [11], and Feidt et al. [12].

Terminology innovations included the introduction of the term "endoreversible" by Rubin [13] to describe the reversibility of the work producing compartment or alternatively the term "exoirreversible" for the external irreversibilities that surrounded the same compartment was mentioned by Radcenco [14]. It is to be noted that the concept of internal reversibility or external irreversibility is a well-established thermodynamic concept. It is tantamount to the local thermodynamic equilibrium model [15] that serves as the foundation for modern heat transfer and fluid dynamics [16]. The term "finite-time" thermodynamics was introduced by Andresen et al. [17] to describe the optimization of thermodynamic processes subject to time constraints [18].

Again, in involvement with actual engines we are concerned with the following factors. The working spaces of many energy-conversion machines operate under the conditions of oscillating flow. These machines include Stirling engines and refrigerators, reciprocating internal combustion engines, gas and refrigerant compressors, cryocooler and expanders, and compressors and pulse-tube refrigerators. Newton's law of cooling as a basis of typical convective heat transfer correlations states that heat transfer is proportional to the bulk gas and wall temperature difference. Fourier's law of conduction, an exact expression within the continuum hypothesis states that heat transfer is proportional to the temperature gradient at the wall. In most steady-state convective heat transfer situations, the wall temperature gradient is proportional to the bulk gas wall temperature difference and so Newton's law works. It is often neglected that Newton's law is an engineering approximation and heat transfer is not necessarily proportional to the bulk gas wall temperature difference. In particular, Newton's law in its ordinary form is not valid in most oscillating pressure or oscillating flow heat transfer. There is a phase shift between heat transfer and temperature difference. So there are sections of the cycle where the conventional convective heat transfer coefficient becomes negative and points where it becomes infinite. In several studies beginning with Gutkowicz-Krusin et al. [19], the assumption that the heat transfer rates are proportional to the local temperature difference were replaced by more general nonlinear heat transfer models that account for natural convection, radiation, and temperature-dependent properties by De Vos [20], Chen and Yan [21], Angulo-Brown and Páez-Hernández [22]. Early studies were also conducted by Rubin [13], Lucca [23], Rozonoer and Tsirlin [24], Mozurkewich and Berry [25], and Tsirlin [26]. In circuit design with alternating electricity, circuit elements are assigned a complex impedance rather than a real resistance. In mechanical vibration analysis, machine elements are assigned complex impedances rather than real inertias, damping constants, and spring constants. Periodic conduction heat transfer problems and other similar differential equations are often solved using the method of complex field. This in essence reduces a periodic time-dependent problem in real variables to a steady-state problem in complex variables. Once the complex solution is found, the results are usually given in terms of heat transfer magnitude and phase. The same information however, can be given in terms of real and imaginary parts of heat transfer. Expressing the thermal resistance of the boundary layer in terms of complex number is a relatively new idea [27–29].

Next, the maximization of work output as opposed to power output was pursued by Grazzini and Gori [30] and Wu et al. [11]. The subtle differences between the maximum power in time-dependent cyclic versus steady flow power plant models were clarified by Kiang and Wu [31]. As a figure of merit in power plant optimization, Angulo-Brown [32] proposed to maximize the so-called ecological function $\dot{W} - T_L\dot{S}_{gen}$ where $\dot{W}$ is the work output rate, $\dot{S}_{gen}$ is the entropy generation rate of the power plant, and T_L is the heat sink temperature. Since in cases ecological function may assume negative values, recently Ust et al. [33] proposed to maximize the quantity $\frac{\dot{W}}{T_L\dot{S}_{gen}}$ known as ecological coefficient of performance.

Also, several of these studies emphasized the importance of matching the temperature of the working fluid to the temperature of the heating agent. We can maximize the instantaneous power output of the model in two ways, with respect to the temperature range spanned by the working fluid (τ) and/or the allocation of the total thermal conductance (x). The practical implications of optimal allocation of heat exchanger inventory x_{opt} are clear and immediate—the heat exchanger inventory must be divided in a certain way. The physical interpretations of optimized temperature range of the working fluid τ_{opt} are more abstract. The message to the designer is that the working fluid must be selected in such a way that it can be heated while at a certain temperature and cooled at another optimal temperature for each given pair of heat exchanger inventories. The designer is considerably less free to experiment with the fluid type than to divide heat exchanger inventory. Another fact of the matter is that large-scale power plants are optimized for fixed heat input and not variable heat input [34].

The objective of the present contribution is to provide an analytical model of a generalized power plant operating cyclically. In this study, both bypass heat leak and internal irreversibility are considered. Unsteady state heat transfer processes are modeled as complex. The conventional Newton's law of cooling is replaced by a generalized power law. The exponent of the power law is also considered to be complex to include the relaxation process in heat transfer of the system [35–37]. Finally, optimization of power output [38, 39] is carried out with respect to the optimal allocation of heat exchanger inventories [40–42] alone. The principle of operation of power plant at maximum power output over other objectives turned out to be the most natural choices as Odum and Pinkerton [43] furnished several examples of this category from the fields of engineering, physics, and biology. The present emphasis examines the result obtained by the author [44] in the context of the law of motive force proposed in this monograph.

6.2 The Physical Model

With reference to Fig. 6.1, we discuss the following modeling features in succession. There remain many engineering problems, for example, heat transfer in combustion engine wall and space reentry problems in which boundary condition functions are time dependent. In nuclear reactor fuel elements during power transients, the energy generation rate varies with time. Duhamel's theorem [45] provides a convenient approach for developing solution to the heat conduction problems with time-dependent boundary conditions and/or time-dependent energy generation. Thus, the external irreversibilities in heat transfer occurring at the hot end and cold end heat exchangers between the heat engine and the corresponding thermal reservoir, as considered by Curzon and Ahlborn [2], can be further modified from a more practical view of modeling. On account of periodic heat transfer mechanism, the temperatures of the heat source (T_H) and the heat sink (T_L) are different from the time-averaged temperatures of the working fluid at the hot end (T_{HO}) and cold side (T_{LO}), respectively.

Thus, the actual temperatures of the working fluid at the hot end and cold side can be considered as the superposition of the steady part with the periodic part. So in transient form, the temperatures of the working fluid at the hot end (T_{HC}) and cold end (T_{LC}) can be expressed in complex combinations as

$$T_{HC} = T_{HO} + T_h \exp(i\omega t) \tag{6.1}$$

and

$$T_{LC} = T_{LO} + T_l \exp(i\omega t) \tag{6.2}$$

where T_h and T_l are some real quantities indicating temperatures, $i = \sqrt{-1}$ and ω is the oscillating periodic frequency. The time-averaged quantities presented in Eqs. (6.1) and (6.2) assume, respectively, $\langle T_{HC} \rangle_t = T_{HO}$ and $\langle T_{LC} \rangle_t = T_{LO}$. It is to be noted that the actual thermal reservoir [46] temperatures are considered to be

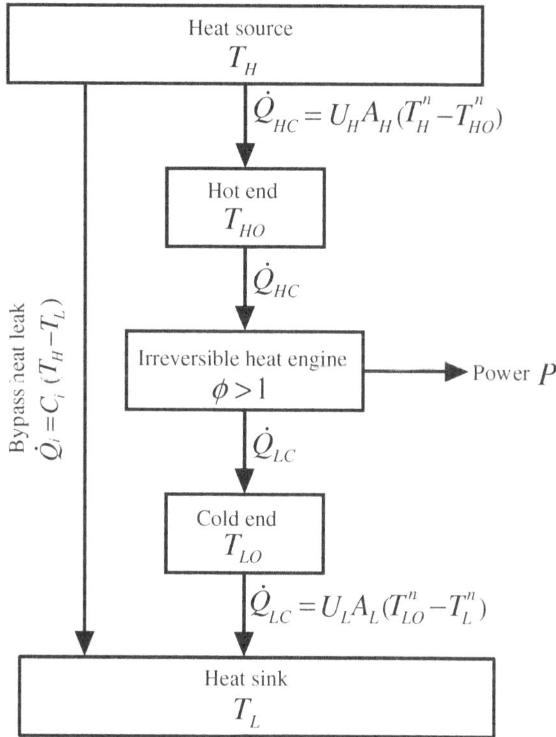

Fig. 6.1 Generalized irreversible heat engine with thermal resistance, bypass heat leak, and internal irreversibility

real and constant in the absence of any periodicity. For the energy flow to occur, the second law of thermodynamics demands that $T_L < T_{LO} < T_{HO} < T_H$.

Next, we relax another stipulation of Curzon-Ahlborn model [2], that is, the heat transfer external to the endoreversible compartment is linear. Instead, we adopt the fact that the unsteady state heat transfer rate $(\dot{Q}')$ between the work producing compartment of the engine and its surroundings follow a generalized power law for the temperature (T) of the form $\dot{Q}' \propto \Delta T^n$ where n is the index of the power law. Thus, for the high temperature side of the heat engine we have

$$\dot{Q}'_{HC} = U_H A_H (T_H^n - T_{HC}^n) \tag{6.3}$$

where U_H is the overall heat transfer coefficient based on heat transfer surface area A_H of the hot end heat exchanger. Similarly, for the low temperature side of the heat engine we obtain

$$\dot{Q}'_{LC} = U_L A_L (T_{LC}^n - T_L^n) \tag{6.4}$$

where U_L is the overall heat transfer coefficient and A_L is the heat transfer surface area of the cold end of the heat exchanger.

Now, we would like to incorporate the physical process of relaxation phenomena [35–37]. One way to incorporate this feature is to consider the index of heat transfer power law to be complex, that is, $n = n_1 + in_2$, where n_1 and n_2 are both real quantities. The imaginary part corresponding to n_2 takes into account the relaxation phenomena in heat transfer mechanism. Also, it is the imaginary part of the index of power law responsible to bring about the temperature to be complex. Further, it is to be noted that for the negative values of the real part of the power law index of heat transfer, overall heat transfer coefficient will have negative values. Such an occurrence is equivalent to the negative heat capacity [47]. Again, in view of steady-state cyclic operation Eqs. (6.3) and (6.4) assume a new form

$$\dot{Q}_{HC} \approx U_H A_H (T_H^n - T_{HO}^n) \tag{6.5}$$

and

$$\dot{Q}_{LC} \approx U_L A_L (T_{LO}^n - T_L^n) \tag{6.6}$$

where the time-averaged quantities of heat transfer rates are $\langle \dot{Q}'_{HC} \rangle_t = \dot{Q}_{HC}$ and $\langle \dot{Q}'_{LC} \rangle_t = \dot{Q}_{LC}$, respectively. The physically realistic basis of approximations incurred in Eqs. (6.5) and (6.6) are, respectively. $\left| \frac{T_h}{T_{HO}} \exp(i\omega t) \right| \ll 1$ or simply $\frac{T_h}{T_{HO}} \ll 1$ and $\left| \frac{T_l}{T_{LO}} \exp(i\omega t) \right| \ll 1$ or simply $\frac{T_l}{T_{LO}} \ll 1$.

Another type of heat loss is from the wall and is known as wall heat loss. A host of power plant elements fall into this category. For example, the heat lost through the wall of a combustion chamber or boiler house, heat removed by the cooling system of an internal combustion engine, and the streamwise convective heat leak channeled toward room temperature by the counter flow heat exchanger of a regenerative Brayton cycle [48]. This modeling feature of heat loss is known as bypass heat leak and was first pointed out by Bejan and Paynter [7]. For this heat transfer rate, $\dot{Q}_i$, which leaks directly through the machine structures and around the power producing compartment, to be constant, we assume

$$\dot{Q}_i = C_i (T_H - T_L) \tag{6.7}$$

where C_i is the shorthand for the internal thermal conductance of the power plant. Upon energy balance we arrive at the following heat transport equalities:

$$\dot{Q}_H = \dot{Q}_i + \dot{Q}_{HC} \tag{6.8}$$

and

$$\dot{Q}_L = \dot{Q}_i + \dot{Q}_{LC} \tag{6.9}$$

where $\dot{Q}_H$ is the heat transfer rate provided by the hot end thermal reservoir and $\dot{Q}_L$ is the heat transfer rate rejected to the cold end of the thermal reservoir.

In reality, the work producing compartment is also irreversible due to various nonequilibrium events inside the engine that ultimately produces the power. So finally we relax the imposed consideration of endoreversibility. This can be simply achieved by the introduction of a dimensionless factor. From the second law of thermodynamics, an irreversible heat engine will release more heat than its counterpart reversible heat engine. If $\dot{Q}_{LC}$ is the rate of heat flow released by the cold working fluid to the cold side heat exchanger and $\dot{Q}_{LCE}$ is that of the endoreversible heat engine, the degree of irreversibility φ can be defined as

$$\phi = \frac{\dot{Q}_{LC}}{\dot{Q}_{LCE}} \geq 1. \tag{6.10}$$

6.3 The Optimization Method

First, we impose the restriction of endoreversibility for the work-producing compartment. Then the second law of thermodynamics applied for this compartment relates the heat transfer and temperature quantities in the following manner:

$$\frac{\dot{Q}_{HC}}{T_{HO}} - \frac{\dot{Q}_{LCE}}{T_{LO}} = 0. \tag{6.11}$$

In the next step, the above limitation is waived with the aid of degree of irreversibility concept introduced in Eq. (6.10). Thus, Eq. (6.10) in combination with Eq. (6.11) labels the work producing compartment to be irreversible. The corresponding relation between the heat transfer and temperature interaction assumes the transformation

$$\frac{\dot{Q}_{LC}}{\dot{Q}_{HC}} = \frac{\dot{Q}_{LC}}{\dot{Q}_{LCE}} \cdot \frac{\dot{Q}_{LCE}}{\dot{Q}_{HC}} = \phi \frac{T_{LO}}{T_{HO}}. \tag{6.12}$$

Next, the power output P of the engine is dictated by the first law of thermodynamics as

$$P = \dot{Q}_H - \dot{Q}_L. \tag{6.13}$$

There is a reduction of power output in presence of bypass heat leak phenomenon. Introducing Eqs. (6.8) and (6.9) into Eq. (6.13) we obtain

$$P = \dot{Q}_H - \dot{Q}_L = (\dot{Q}_i + \dot{Q}_{HC}) - (\dot{Q}_i + \dot{Q}_{LC}) = \dot{Q}_{HC} - \dot{Q}_{LC}. \tag{6.14}$$

The first law-based thermodynamic efficiency of the heat engine is defined as

$$\eta = \frac{P}{\dot{Q}_H} = \frac{P}{\dot{Q}_i + \dot{Q}_{HC}}. \tag{6.15}$$

Now, in order to limit the degrees of freedom of the physical model for the objective of optimization, we look back into Eqs. (6.5) and (6.6). We propose on the basis of the law of motive force that the proportionality constants compete with each other being the forward and backward motivation. From the physics of the problem, the proportionality constants are the conductances of the respective heat exchangers and being the commodities of short supply, the competition is very obvious. Thus, it makes perfect sense to recognize that the total conductance inventory as a constraint also obeys the relation [42]

$$U_H A_H + U_L A_L = UA \tag{6.16}$$

where U is the overall heat transfer coefficient and A is the total heat exchanger area. Accounting for such a constraint of the type Eq. (6.16) is recognized as the law of motive force proposed in this monograph. In Eq. (6.16), the right side is a constant and the left side components are forward and backward motivation with respect to each other. Among other choices such as area constraint [42], cost constraint [49], and the entropy generation rate constraint [50], Eq. (6.16) is indicated to be the most natural selection [51]. Henceforth, we consider the conductances to be the single entity of the proportionality constants rather than the products of overall heat transfer coefficients and corresponding areas in Eqs. (6.5) and (6.6). Mathematically, we mean $(UA)_H = U_H A_H$ and $(UA)_L = U_L A_L$, where UA itself stands for a single commodity. In terms of conductance allocation ratio x, we may write

$$(UA)_H = \left(\frac{x}{1+x}\right) UA,\ (UA)_L = \left(\frac{1}{1+x}\right) UA,\ \text{and}\ \frac{(UA)_L}{(UA)_H} = \frac{1}{x}. \tag{6.17}$$

Further, we define various intermediate temperature ratios as

$$\tau = \frac{T_L}{T_H},\ \tau_o = \frac{T_{LO}}{T_{HO}},\ \text{and}\ \tau_h = \frac{T_{HO}}{T_H}. \tag{6.18}$$

Now, from Eq. (6.12) employing Eqs. (6.5) and (6.6) and invoking nondimensional parameters defined in Eqs. (6.17) and (6.18), we obtain

$$\tau_h^n = \frac{\phi x \tau_o^{1-n} + \left(\frac{\tau}{\tau_o}\right)^n}{1 + \phi x \tau_o^{1-n}}. \tag{6.19}$$

Similarly, with the aid of Eqs. (6.12), (6.5), and (6.6), Eq. (6.14) for the power output transforms into

$$\bar{P} = (1 - \phi\tau_o)\left(\frac{x}{1+x}\right)(1 - \tau_h^n) \quad (6.20)$$

where the dimensionless power $\bar{P}$ is defined as $\bar{P} = \frac{P}{UAT_H^n}$. The elimination of τ_h term between Eqs. (6.19) and (6.20) leads to the result

$$\bar{P} = (1 - \phi\tau_o)\left(\frac{x}{1+x}\right)\left[\frac{1 - \left(\frac{\tau}{\tau_o}\right)^n}{1 + \phi x \tau_o^{1-n}}\right]. \quad (6.21)$$

Equation (6.21) can be rearranged to obtain

$$\left(\frac{1 - \phi\tau_o}{\bar{P}}\right)\left(\frac{x}{1+x}\right) = \frac{1 + \phi x \tau_o^{1-n}}{1 - \left(\frac{\tau}{\tau_o}\right)^n}. \quad (6.22)$$

Now, deploying Eqs. (6.5) and (6.21) into Eq. (6.15), we obtain a revised expression for the efficiency as

$$\eta = \frac{1 - \phi\tau_o}{1 + \bar{\dot{Q}}_i\left[\frac{1 + \phi x \tau_o^{1-n}}{1 - \left(\frac{\tau}{\tau_o}\right)^n}\right]} \quad (6.23)$$

where the dimensionless bypass heat leak is denoted by $\bar{\dot{Q}}_i = \frac{\dot{Q}}{UAT_H^n}$.

Invoking the intermediate relation (6.22) into Eq. (6.23) we arrive at the more simplified expression for efficiency as

$$\eta = \frac{1 - \phi\tau_o}{1 + \bar{\dot{Q}}_i\left(\frac{x}{1+x}\right)\left(\frac{1 - \phi\tau_o}{\bar{P}}\right)}. \quad (6.24)$$

A simple arrangement of Eq. (6.24) leads to the expression

$$\bar{\dot{Q}}_i\left(\frac{x}{1+x}\right)\frac{1}{\bar{P}} - \frac{1}{\eta} = \frac{1}{\phi\tau_o - 1}. \quad (6.25)$$

Equation (6.25) is an interesting result in thermodynamic optimization. When the parameters contained in this equation do not enjoy any degrees of freedom, we can find either the optimal efficiency for a given power output or the maximum power output for a given efficiency. The result found from these two extremal conditions are the same, so either of the two conditions may be used [22]. In a variational formulation of the problem of this type [21], the situation represents a transversality condition [52]. From the physics of the relation, Eq. (6.25) represents a competition between the power output and the corresponding efficiency of the engine. Thus, in another sense Eq. (6.25) is a variant of the law of motive force introduced as an optimization philosophy in this treatise.

Since index n of the power law is indicated to be a complex quantity, the expression for the power output obtained from Eq. (6.21) represents a complex number. However, the magnitude of the power output is the real part of that complex number. So we proceed to extract the real part of the power in the following manner. If b is a nonzero complex number, and assuming a is a completely arbitrary complex number, then by the general power b^a we mean every value given by the formula $b^a = e^{a \ln b}$ [53]. Thus, we have for

$$\left(\frac{\tau}{\tau_o}\right)^n = \left(\frac{\tau}{\tau_o}\right)^{n_1+in_2} = \left(\frac{\tau}{\tau_o}\right)^{n_1} \cos\left[n_2 \ln\left(\frac{\tau}{\tau_o}\right)\right] + i\left(\frac{\tau}{\tau_o}\right)^{n_1} \sin\left[n_2 \ln\left(\frac{\tau}{\tau_o}\right)\right] \quad (6.26)$$

and

$$\tau_o^{1-n} = \tau_o^{(1-n_1)-in_2} = \tau_o^{1-n_1} \cos[n_2 \ln(\tau_o)] - i\tau_o^{1-n_1} \sin[n_2 \ln(\tau_o)]. \quad (6.27)$$

So the expression contributing to the complex number in Eq. (6.21) is expressed as

$$\frac{1-\left(\frac{\tau}{\tau_o}\right)^n}{1+\phi x \tau_o^{1-n}} = \frac{c_1 - ic_2}{c_4 - ic_6} = \left(\frac{c_1c_4 + c_2c_6}{c_4^2 + c_6^2}\right) + i\left(\frac{c_1c_6 - c_2c_4}{c_4^2 + c_6^2}\right) \quad (6.28)$$

where

$$c_1 = 1 - \left(\frac{\tau}{\tau_o}\right)^{n_1} \cos\left[n_2 \ln\left(\frac{\tau}{\tau_o}\right)\right], \quad (6.29)$$

$$c_2 = \left(\frac{\tau}{\tau_o}\right)^{n_1} \sin\left[n_2 \ln\left(\frac{\tau}{\tau_o}\right)\right], \quad (6.30)$$

$$c_3 = \phi\tau_o^{1-n_1} \cos[n_2 \ln(\tau_o)], \quad (6.31)$$

$$c_4 = 1 + c_3 x, \quad (6.32)$$

$$c_5 = \phi\tau_o^{1-n_1} \sin[n_2 \ln(\tau_o)], \quad (6.33)$$

and

$$c_6 = c_5 x. \quad (6.34)$$

Thus, the real part of the power delivered assumes the form

$$P_r = \mathrm{Re}(\bar{P}) = (1 - \phi\tau_o)\left(\frac{x}{1+x}\right)\left(\frac{c_7 x + c_1}{c_8 x^2 + 2c_3 x + 1}\right) \quad (6.35)$$

where P_r stands for the real part of the power and the parameters are

$$c_7 = c_1 c_3 + c_2 c_5 \tag{6.36}$$

and

$$c_8 = c_3^2 + c_5^2. \tag{6.37}$$

Equation (6.35) may further be rearranged as

$$P_r = (1 - \phi\tau_o)\left(\frac{c_7 x^2 + c_1 x}{c_8 x^3 + c_9 x^2 + c_{10} x + 1}\right) \tag{6.38}$$

where

$$c_9 = 2c_3 + c_8 \tag{6.39}$$

and

$$c_{10} = 2c_3 + 1. \tag{6.40}$$

Now, we return to our fundamental objective of optimal heat exchanger allocation for which power output is maximum. Thus, setting the first derivative equal to zero in Eq. (6.38) with respect to the optimal heat exchanger allocation ratio x, we obtain

$$x^4 + c_{11} x^3 + c_{12} x^2 + c_{13} x + c_{14} = 0 \tag{6.41}$$

where

$$c_{11} = 2\frac{c_1}{c_7}, \tag{6.42}$$

$$c_{12} = \frac{c_1 c_9 - c_7 c_{10}}{c_7 c_8}, \tag{6.43}$$

$$c_{13} = -2\frac{1}{c_8}, \tag{6.44}$$

and

$$c_{14} = -\frac{c_1}{c_7 c_8}. \tag{6.45}$$

The quartic [54] equation (6.41) is a commonplace occurrence in many physical problems and can be solved completely analytically in the following steps. The algebraic form in which the solutions of the quartic equation normally appear is so awkward and clumsy that they are of little use for further manipulations in obtaining closed-form expressions of other related quantities. Thus, we proceed to furnish an elegant expression for the optimal heat exchanger allocation ratio. First, we reduce the quartic equation into a cubic equation on following Descartes' method [55]. Substituting $x = y - \frac{1}{4}c_{11}$ into Eq. (6.41) we have the transformed equation as

$$y^4 + c_{15}y^2 + c_{16}y + c_{17} = 0 \tag{6.46}$$

where

$$c_{15} = c_{12} - \frac{3}{8}c_{11}^2, \tag{6.47}$$

$$c_{16} = c_{13} + \frac{1}{8}c_{11}^3 - \frac{1}{2}c_{11}c_{12}, \tag{6.48}$$

and

$$c_{17} = c_{14} - \frac{3}{256}c_{11}^4 + \frac{1}{16}c_{11}^2c_{12} - \frac{1}{4}c_{11}c_{13}. \tag{6.49}$$

Now, the solution of the transformed quartic equation (6.46) can be obtained from the following auxiliary cubic equation in terms of the variable z as:

$$z^3 + c_{18}z^2 + c_{19}z + c_{20} = 0 \tag{6.50}$$

where

$$c_{18} = \frac{1}{2}c_{15}, \tag{6.51}$$

$$c_{19} = \frac{1}{16}\left(c_{15}^2 - 4c_{17}\right), \tag{6.52}$$

and

$$c_{20} = -\frac{1}{64}c_{16}^2. \tag{6.53}$$

If z_1, z_2, and z_3 are the three roots of Eq. (6.50), then the four roots of the transformed quartic equation in y are [55] $\pm\sqrt{z_1} \pm \sqrt{z_2} \pm \sqrt{z_3}$. So the solutions of the original equation in x are

$$x_1 = \sqrt{z_1} + \sqrt{z_2} + \sqrt{z_3} - \frac{1}{4}c_{11}, \tag{6.54}$$

$$x_2 = \sqrt{z_1} - \sqrt{z_2} - \sqrt{z_3} - \frac{1}{4}c_{11}, \tag{6.55}$$

$$x_3 = -\sqrt{z_1} + \sqrt{z_2} - \sqrt{z_3} - \frac{1}{4}c_{11}, \tag{6.56}$$

and

$$x_4 = -\sqrt{z_1} - \sqrt{z_2} + \sqrt{z_3} - \frac{1}{4}c_{11}. \tag{6.57}$$

Next, we are concerned about the solution of the cubic equation (6.50). We adopt the procedure outlined by McKelvey [56]. Substituting $z = \bar{z} - \frac{1}{3}c_{18}$ into Eq. (6.50) we obtain

$$\bar{z}^3 \pm 3c_{21}\bar{z} + 2c_{22} = 0 \tag{6.58}$$

where

$$\pm 3c_{21} = c_{19} - \frac{1}{3}c_{18}^2 \tag{6.59}$$

and

$$2c_{22} = \frac{2}{27}c_{18}^3 - \frac{1}{3}c_{18}c_{19} + c_{20}. \tag{6.60}$$

Then the roots of Eq. (6.50) can be listed as follows [56]:

Case I: When $c_{21} > 0$ one real root and two complex roots are

$$z_1 = -2\sqrt{c_{21}}\sinh\left[\frac{1}{3}\sinh^{-1}\left(\frac{c_{22}}{c_{21}^{3/2}}\right)\right] - \frac{1}{3}c_{18}, \tag{6.61}$$

$$z_2 = \sqrt{c_{21}}\left\{\sinh\left[\frac{1}{3}\sinh^{-1}\left(\frac{c_{22}}{c_{21}^{3/2}}\right)\right] + i\sqrt{3}\cosh\left[\frac{1}{3}\sinh^{-1}\left(\frac{c_{22}}{c_{21}^{3/2}}\right)\right]\right\} - \frac{1}{3}c_{18}, \tag{6.62}$$

and

$$z_3 = \sqrt{c_{21}}\left\{\sinh\left[\frac{1}{3}\sinh^{-1}\left(\frac{c_{22}}{c_{21}^{3/2}}\right)\right] - i\sqrt{3}\cosh\left[\frac{1}{3}\sinh^{-1}\left(\frac{c_{22}}{c_{21}^{3/2}}\right)\right]\right\} - \frac{1}{3}c_{18}. \tag{6.63}$$

Case II: When $c_{21} < 0$ and $c_{22}^2 - c_{21}^3 > 0$ one real root and two complex roots are

$$z_1 = -2\sqrt{c_{21}}\cosh\left[\frac{1}{3}\cosh^{-1}\left(\frac{c_{22}}{c_{21}^{3/2}}\right)\right] - \frac{1}{3}c_{18}, \tag{6.64}$$

$$z_2 = \sqrt{c_{21}}\left\{\cosh\left[\frac{1}{3}\cosh^{-1}\left(\frac{c_{22}}{c_{21}^{3/2}}\right)\right] + i\sqrt{3}\sinh\left[\frac{1}{3}\cosh^{-1}\left(\frac{c_{22}}{c_{21}^{3/2}}\right)\right]\right\} - \frac{1}{3}c_{18}, \tag{6.65}$$

and

$$z_3 = \sqrt{c_{21}}\left\{\cosh\left[\frac{1}{3}\cosh^{-1}\left(\frac{c_{22}}{c_{21}^{3/2}}\right)\right] - i\sqrt{3}\sinh\left[\frac{1}{3}\cosh^{-1}\left(\frac{c_{22}}{c_{21}^{3/2}}\right)\right]\right\} - \frac{1}{3}c_{18}. \tag{6.66}$$

Case III: When $c_{21} < 0$ and $c_{22}^2 - c_{21}^3 < 0$ three real roots are

$$z_1 = -2\sqrt{c_{21}}\cos\left[\frac{1}{3}\cos^{-1}\left(\frac{c_{22}}{c_{21}^{3/2}}\right)\right] - \frac{1}{3}c_{18}, \tag{6.67}$$

$$z_2 = \sqrt{c_{21}}\left\{\cos\left[\frac{1}{3}\cos^{-1}\left(\frac{c_{22}}{c_{21}^{3/2}}\right)\right] + \sqrt{3}\sin\left[\frac{1}{3}\cos^{-1}\left(\frac{c_{22}}{c_{21}^{3/2}}\right)\right]\right\} - \frac{1}{3}c_{18}, \tag{6.68}$$

and

$$z_3 = \sqrt{c_{21}}\left\{\cos\left[\frac{1}{3}\cos^{-1}\left(\frac{c_{22}}{c_{21}^{3/2}}\right)\right] - \sqrt{3}\sin\left[\frac{1}{3}\cos^{-1}\left(\frac{c_{22}}{c_{21}^{3/2}}\right)\right]\right\} - \frac{1}{3}c_{18}. \tag{6.69}$$

In order to calculate the roots of Eq. (6.41) via Eqs. (6.61) through (6.69), it is convenient to employ the following relations [57]:

$$\sinh^{-1}\left(\frac{c_{22}}{c_{21}^{3/2}}\right) = \ln\left[\frac{c_{22}}{c_{21}^{3/2}} + \sqrt{\left(\frac{c_{22}}{c_{21}^{3/2}}\right)^2 + 1}\right] \tag{6.70}$$

and

$$\cosh^{-1}\left(\frac{c_{22}}{c_{21}^{3/2}}\right) = \pm\ln\left[\frac{c_{22}}{c_{21}^{3/2}} + \sqrt{\left(\frac{c_{22}}{c_{21}^{3/2}}\right)^2 - 1}\right]. \tag{6.71}$$

Equation (6.70) is numerically robust for $\frac{c_{22}}{c_{21}^{3/2}} \geq 0$. For negative values of $\frac{c_{22}}{c_{21}^{3/2}}$ we utilize the symmetry property $\sinh^{-1}\left(-\frac{c_{22}}{c_{21}^{3/2}}\right) = -\sinh^{-1}\left(\frac{c_{22}}{c_{21}^{3/2}}\right)$. It is to be noted that Eq. (6.71) is valid only for $\frac{c_{22}}{c_{21}^{3/2}} \geq 1$.

Table 6.1 Effect of power law on the heat exchanger allocation and the thermal efficiency of the engine

Sl. No.	τ	τ_o	n_1	n_2	ϕ	$\bar{\dot{Q}}_i$	x	η_r
1	$\frac{2}{5}$	$\frac{2}{3}$	1.0	0.0	1.0	0.025	1.0	0.296296269
2	$\frac{2}{5}$	$\frac{2}{3}$	−1.0	0.0	1.0	0.025	1.50000012	0.355555534
3	$\frac{2}{5}$	$\frac{2}{3}$	4.0	0.0	1.0	0.025	0.544331133	0.308217049

Thus, we obtain an exact analytical expression for the real part of the power from Eq. (6.35). Then the real part of the efficiency is calculated from Eq. (6.25) using Eq. (6.35) as

$$\eta_r = \mathrm{Re}(\eta) = \frac{1}{\frac{1}{1-\phi\tau_o} + \dot{Q}_i\left(\frac{x}{1+x}\right)\frac{1}{P_r}} \tag{6.72}$$

where η_r is the real part of the efficiency.

6.4 Numerical Examples

The analytical results obtained in the above section can be utilized to quantify the effects of various parameters on the allocation of heat exchanger inventory and the thermal efficiency of a real heat engine. In performing the numerical analysis of the physical model presented here, we adopt closely the following data from the existing power plants available in the open literature [58–60].

From the physical point of view, the temperatures of the heat source and sink are considered not controllable. Similarly, the temperatures of the hot end and cold side of the working fluid are also considered to be fixed for practical reasons. Thus, in this parametric study we assume $\tau = \frac{2}{5}$ and $\tau_o = \frac{2}{3}$ not to vary. The discrete variations of other parameters considered are as follows: $n_1 = 1.0, -1.0$, and 4.0; $n_2 = 0.0, 0.005, 0.025$, and 0.125; $\phi = 1.0, 1.005, 1.010$, and 1.015 and $\bar{\dot{Q}}_i =$ 0.001, 0.005, 0.025, and 0.125. It is to be remarked that $n_1 = 1.0$ represents conventional Newton's law of convective cooling, whereas $n_1 = -1.0$ demonstrates the phenomenological heat transfer and $n_1 = 4.0$ categorizes the radiative heat transfer mode.

Table 6.1 shows the alteration of heat exchanger allocation ratio and the efficiency with the variation of the real part of the power law index n_1 when the other parameters assume some representative values. It is observed that n_1 has a major impact both on the heat exchanger allocation ratio and on the thermal efficiency of the power plant. It is interesting to report that optimal heat exchanger allocation ratio conforms to the macroscopic organization with equipartition principle when the law of heat resistance is linear, that is, for $n_1 = 1.0$. On the other hand, for the

Table 6.2 Effect of relaxation on the heat exchanger allocation and the thermal efficiency of the engine

Sl. No.	τ	τ_o	n_1	n_2	ϕ	$\overline{\dot{Q}_i}$	x	η_r
1	$\frac{2}{5}$	$\frac{2}{3}$	1.0	0.005	1.0	0.025	0.997458279	0.296254754
2	$\frac{2}{5}$	$\frac{2}{3}$	1.0	0.025	1.0	0.025	0.987142146	0.296092391
3	$\frac{2}{5}$	$\frac{2}{3}$	1.0	0.125	1.0	0.025	0.932174623	0.295382202

Table 6.3 Effect of internal irreversibility on the heat exchanger allocation and the thermal efficiency of the engine

Sl. No.	τ	τ_o	n_1	n_2	ϕ	$\overline{\dot{Q}_i}$	x	η_r
1	$\frac{2}{5}$	$\frac{2}{3}$	1.0	0.025	1.005	0.025	0.984683096	0.293090314
2	$\frac{2}{5}$	$\frac{2}{3}$	1.0	0.025	1.010	0.025	0.982242286	0.290089190
3	$\frac{2}{5}$	$\frac{2}{3}$	1.0	0.025	1.015	0.025	0.979819536	0.287089020

Table 6.4 Effect of bypass heat leak on the heat exchanger allocation and the thermal efficiency of the engine

Sl. No.	τ	τ_o	n_1	n_2	ϕ	$\overline{\dot{Q}_i}$	x	η_r
1	$\frac{2}{5}$	$\frac{2}{3}$	1.0	0.025	1.0	0.001	0.987142146	0.331664711
2	$\frac{2}{5}$	$\frac{2}{3}$	1.0	0.025	1.0	0.005	0.987142146	0.325154096
3	$\frac{2}{5}$	$\frac{2}{3}$	1.0	0.025	1.0	0.125	0.987142146	0.204640388

nonlinear laws of heat transfer, the principle of equipartition does not hold any longer. For the phenomenological heat transfer law, that is, for $n_1 = -1.0$, the hot end side has almost 60 % and the cold end has nearly 40 % of the total heat exchanger inventory. Again, for the radiative mode of heat transfer law, that is, for $n_1 = 4.0$, we allocate 35 % to the hot side and 65 % to the cold side out of the total heat exchanger inventory available. Also, it is to be noticed that there is a gain in the efficiency of the engine in departing from the linear law of external heat transfer resistance with a subsequent deviation from the equipartitioned allocation of heat exchanger inventory.

Table 6.2 represents the effect of relaxation in heat transfer both on the heat exchanger allocation and the efficiency of the engine. It is observed that when the parameter n_2 is high indicting a prominent effect of relaxation, the heat exchanger allocation ratio departs from the equipartition principle and there is a subsequent drop in efficiency of the engine. The higher the magnitude of relaxation effect in heat transfer, the greater is the deviation from equal allocation of heat exchanger, and the lesser is the thermal efficiency.

Table 6.3 reveals the effect of irreversibility of the work producing compartment. It is seen that within the range of variation of the parameter ϕ considered, the optimal heat exchanger allocation ratio drops slightly to nearly follow the equipartition principle in macroscopic organization and in correspondence there remains a little sparing of the thermal efficiency of the heat engine.

Table 6.4 demonstrates the influence of bypass heat leak on the heat exchanger allocation ratio and the thermal efficiency of the engine. It is found that bypass heat leak is not sensitive to the optimal heat exchanger allocation ratio but the thermal efficiency of the engine is affected. Optimal heat exchanger inventory distribution closely follows the macroscopic organization with the principle of equipartition. The stronger the effect of bypass heat leak, the greater the drop in the thermal efficiency of the heat engine.

6.5 Discussions

In this chapter, an analytical model of a Carnot-like heat engine in the presence of power law, external thermal resistance, relaxation effect in heat transfer, bypass heat leak, and internal irreversibility is presented. The thermal efficiency of the heat engine with the objective function of maximum power output was investigated under the influence of various parameters. Following the law of motive force, it is observed that the power output and thermal efficiency of the heat engine competes with each other and thus supports a body of observations in the theory of finite-time thermodynamics [21, 61]. This competition is also the very philosophy of the law of motive force proposed in this monograph. For practical reasons, the true engineering quest of optimal heat exchanger allocation for maximum power output is attended, leaving behind the influence of optimal intermediate temperature ratio of the hot end and the cold side of the working fluid and also the effect of heat sink as well as heat source temperature ratio. It is noted that for optimal allocation of the heat exchanger inventory, which maintains a category of macroscopic organization with the principle of equipartition, the efficiency at maximum power output also tends to assume a representative value [62]. The optimal heat exchanger allocation and the maximum power efficiency are both drastically affected by the selection of power law for the external heat transfer resistance. For a choice of radiative mode of power law for the external heat transfer, there is a great deviation from equipartitioned allocation of heat exchanger equipment, but with a subsequent increase in efficiency at maximum power. For an enhanced effect of relaxation phenomenon in heat transfer the optimal heat exchanger allocation ratio deviates relatively to a small extent and induces a negligible drop in the maximum power efficiency. The actual effect of internal irreversibility is not very prominent on the heat exchanger allocation ratio and the maximum power efficiency. Thus, endoreversibility is a concrete concept in the finite-time thermodynamic formulation of thermal systems. The bypass heat leak being diffused through the mechanical support only is considered to be linear and thus renders a noticeable effect on the maximum power efficiency for a relatively high value of the concerned parameter leaving the optimal heat exchanger allocation ratio to be near the equipartitioned value.

References

1. Moran, M.J.: On the second-law analysis and the failed promise of finite-time thermodynamics. Energy **23**, 517–519 (1998)
2. Curzon, F.L., Ahlborn, B.: Efficiency of a Carnot engine at maximum power output. Am. J. Phys. **43**, 22–24 (1975)
3. Carnot, S.: Reflections on the motive power of fire, and on machines fitted to develop that power (trans: Thurston, R.H.). In: Mendoza, E. (ed.) Reflections on the Motive Power of Fire. Dover, New York (2005)
4. Clapeyron, É.: Memoir on the motive power of heat (trans: Mendoza, E.). In: Mendoza, E. (ed.) Reflections on the Motive Power of Fire. Dover, New York (2005)
5. Landsberg, P.T., Leff, H.S.: Thermodynamic cycles with nearly universal maximum-work efficiencies. J. Phys. A Math. Gen. **22**, 4019–4026 (1989)
6. Novikov, I.I.: The efficiency of atomic power stations. J. Nucl. Energy II **7**, 125–128 (1958)
7. Bejan, A., Paynter, H.M.: Solved Problems in Thermodynamics. Problem VIID. MIT, Cambridge (1976)
8. Andresen, B., Salamon, P., Berry, R.S.: Thermodynamics in finite-time: extremals for imperfect heat engines. J. Chem. Phys. **66**, 1571–1577 (1977)
9. Lu, P.C.: On optimal disposal of waste heat. Energy **5**, 993–998 (1980)
10. Andresen, B., Salamon, P., Berry, R.S.: Thermodynamics in finite time. Phys. Today **37**, 62–70 (1984)
11. Wu, C., Kiang, R.L., Lopardo, V.J., Karpouzian, G.N.: Finite-time thermodynamics and endoreversible heat engines. Int. J. Mech. Eng. Ed. **21**, 337–346 (1993)
12. Feidt, M., Ramany-Bala, P., Benelmir, R.: Synthesis, unification and generalization of preceding finite time thermodynamic studies devoted to Carnot cycles. In: Carnevalle, E., Manfrida, G., Martelli, F. (eds.) Proceedings of the Florence World Energy Research Symposium (FLOWERS'94), SGEditoriali, Padova (1994)
13. Rubin, M.H.: Optimal configuration of a class of irreversible heat engines-I. Phys. Rev. A **19**, 1272–1276 (1977)
14. Radcenco, V.: Definition of exergy, heat and mass fluxes based on the principle of minimum entropy generation. Rev. Chim. **41**, 40–49 (1990) (in Romanian)
15. Bejan, A.: Advanced Engineering Thermodynamics, pp. 62–64. Wiley, New York (2006)
16. Berg, C.: Conceptual issues in energy efficiency. In: Valero, A., Tsatsaronis, G. (eds.) Proceedings of the International Symposium on Efficiency, Costs, Optimization and Simulation of Energy Systems (ECOS'92), Zaragoza (1992)
17. Andresen, B., Berry, R.S., Nitzan, A., Salamon, P.: Thermodynamics in finite time-I. The step-Carnot cycle. Phys. Rev. A **15**, 2086–2093 (1977)
18. Tolman, R.C., Fine, P.C.: On the irreversible production of entropy. Rev. Mod. Phys. **20**, 51–77 (1948)
19. Gutkowicz-Krusin, D., Procaccia, I., Ross, J.: On the efficiency of rate processes. Power and efficiency of heat engines. J. Chem. Phys. **69**, 3898–3906 (1978)
20. De Vos, A.: Efficiency of some heat engines at maximum-power conditions. Am. J. Phys. **53**, 570–573 (1985)
21. Chen, L., Yan, Z.: The effect of heat-transfer law on performance of a two-heat-source endoreversible cycle. J. Chem. Phys. **90**, 3740–3743 (1989)

22. Angulo-Brown, F., Páez-Hernández, R.: Endoreversible thermal cycle with a nonlinear heat transfer law. J. Appl. Phys. **74**, 2216–2219 (1993)
23. Lucca, G.: Un corollario al teorema di Carnot. Cond. Aria Risc. Refrig., pp. 19–42, January 1981 (in Italian)
24. Rozonoer, L.I., Tsirlin, A.M.: Optimal control of thermodynamic processes-III. Autom. Remote Control **44**, 314–326 (1983)
25. Mozurkewich, M., Berry, R.S.: Optimization of a heat engine based on a dissipative system. J. Appl. Phys. **54**, 3651–3661 (1983)
26. Tsirlin, A.M.: Optimal control of irreversible heat and mass transfer processes. Tekh. Kybernet. **2**, 171–177 (1991) (in Russian)
27. Pfreim, H.: Beitrag zur Theorie der Wärmeleitung bei periodisch veränderlichen (quasistationären) Temperaturfeldern. Ing. Arch. **6**, 97–127 (1935) (in German)
28. Pfreim, H.: Der periodische Wärmeübergang bei kleinen Druckschwankungen. Forsch. Ing. Wes. **11**, 67–75 (1940) (in German)
29. Pfreim, H.: Probleme der periodischen Wärmeübertragung mit Bezug auf Kolbenmaschinen. Forsch. Ing. Wes. **10**, 27–40 (1939) (in German)
30. Grazzini, G., Gori, F.: Influence of thermal irreversibilities on work producing systems. Rev. Gen. Therm. **312**, 637–639 (1987)
31. Kiang, R.L., Wu, C.: Clarification of finite-time thermodynamic cycle analysis. Int. J. Power Energy Syst. **14**, 68–71 (1994)
32. Angulo-Brown, F.: An ecological optimization criterion for finite-time heat engines. J. Appl. Phys. **69**, 7465–7469 (1991)
33. Ust, Y., Sahin, B., Kodal, A.: Performance analysis of an irreversible Brayton heat engine based on ecological coefficient of performance criterion. Int. J. Therm. Sci. **45**, 94–101 (2006)
34. Bejan, A.: Entropy Generation Minimization, p. 210. CRC Press, New York (1996)
35. Kornhauser, A.A., Smith Jr., J.L.: Application of a complex Nusselt-number to heat transfers during compression and expansion. J. Heat Transf. **116**, 536–542 (1994)
36. Meixner, J.: Thermodynamic theory of relaxation phenomena. In: Donnelly, R.J., Herman, R., Prigogine, I. (eds.) Non-Equilibrium Thermodynamics, Variational Techniques and Stability. Chicago University Press, Chicago (1966)
37. Wu, F., Wu, C., Guo, F., Li, Q., Chen, L.: Optimization of a thermoacoustic engine with a complex heat transfer exponent. Entropy **5**, 444–451 (2003)
38. De Mey, G., De Vos, A.: On the optimum efficiency of endoreversible thermodynamic processes. J. Phys. D Appl. Phys. **27**, 736–739 (1994)
39. Gordon, J.M., Huleihil, M.: On optimizing maximum-power heat engines. J. Appl. Phys. **69**, 1–7 (1991)
40. Bejan, A.: Power and refrigeration plants for minimum heat exchanger size. J. Energy Res. Technol. **115**, 148–150 (1993)
41. Bejan, A.: Theory of heat transfer-irreversible power plants-II. The optimal allocation of heat exchange equipment. Int. J. Heat Mass Transf. **38**, 433–444 (1995)
42. Burzler, J.M., Amelkin, S.A., Tsirlin, A.M., Hoffmann, K.H.: Optimal allocation of heat exchanger investment. Open Sys. Inf. Dyn. **11**, 291–306 (2004)
43. Odum, H.T., Pinkerton, R.C.: Time's speed regulator: the optimum efficiency for maximum power output in physical and biological systems. Am. Sci. **43**, 331–343 (1955)
44. Pramanick, A.K., Das, P.K.: Assessment of Bejan's heat exchanger allocation model under the influence of generalized thermal resistance, relaxation effect, bypass heat leak, and internal irreversibility. Int. J. Heat Mass Transf. **51**, 474–484 (2008)
45. Carslaw, H.S., Jaeger, J.C.: Conduction of Heat in Solids, pp. 30–33. Oxford University Press, Oxford (1959)
46. Attard, P.: Thermodynamics and Statistical Mechanics, pp. 21–23. Academic Press, London (2002)
47. Landsberg, P.T.: Thermodynamics, pp. 42, 45. Wiley-Interscience, New York (1961)

48. Bejan, A.: A general variational principle for thermal insulation system design. Int. J. Heat Mass Transf. **22**, 219–228 (1979)
49. Antar, M.A., Zubair, S.M.: Thermoeconomic considerations in the optimum allocation of heat exchanger inventory for a power plant. Energy Convers. Manage. **42**, 1169–1179 (2001)
50. Ait-Ali, M.: Maximum power and thermal efficiency of an irreversible power cycle. J. Appl. Phys. **78**, 4313–4318 (1995)
51. Pramanick, A.K., Das, P.K.: Constructal design of a thermoelectric device. Int. J. Heat Mass Transf. **49**, 1420–1429 (2006)
52. Gelfand, I.M, Fomin, S.V.: Calculus of Variations (trans: Silverman, R.A.), pp. 59–61. Dover, New York (2000)
53. Knopp, K.: Elements of the Theory of Functions (trans: Bagemihl, F.), p. 134. Dover, New York (1952)
54. Lanczos, C.: Applied Analysis, pp. 19–22. Dover, New York (1988)
55. Dickson, L.E.: Elementary Theory of Equations, pp. 42–44. Wiley, New York (1914)
56. McKelvey, J.P.: Simple transcendental expressions for the roots of cubic equations. Am. J. Phys. **52**, 269–270 (1984)
57. Press, W.H., Teukolsky, S.A., Vetterling, W.T., Flannery, B.P.: Numerical Recipes in Fortran 77, pp. 178–180. Cambridge University Press, Cambridge (2001)
58. Chierici, A.: Planning of a geothermal power plant: technical and economic principles-III. In: U. N. Conference on New Sources of Energy, pp. 299–311, New York (1964)
59. Griffiths, G.M.: CANDU—a Canadian success story. Phys. Can. **30**, 2–6 (1974)
60. Spalding, D.B., Cole, E.H.: Engineering Thermodynamics, p. 209. Edward Arnold, London (1973)
61. Chen, J., Yan, Z., Lin, G., Andresen, B.: On the Curzon-Ahlborn efficiency and its connection with the efficiencies of real heat engines. Energy Convers. Manage. **42**, 173–181 (2001)
62. Bejan, A.: Advanced Engineering Thermodynamics, pp. 377–378. Wiley, New York (1997)

About the Author

Achintya Kumar Pramanick received his Bachelor of Mechanical Engineering degree in July 1993 from National Institute of Technology Durgapur, India, formerly known as Regional Engineering College Durgapur. He obtained his first Master of Mechanical Engineering degree with specialization in Heat Power from Jadavpur University, India in February 1996. He was a permanent faculty in Mechanical Engineering Department at North Eastern Regional Institute of Science and Technology, India from March 1996 to September 1997. He taught at the Department of Mechanical Engineering of Jalpaiguri Government College, India from September 1997 to July 1998 as a permanent faculty. He also served his Alma Mater, National Institute of Technology Durgapur as a permanent faculty from August 1998 to July 2000. He worked at Louisiana State University, Baton Rouge, USA as a teaching and research assistant while pursuing his second Master of Science degree in Mechanical Engineering with specialization in Thermofluid Science between August 2000 to August 2002. He was the recipient of Deutscher Akademischer Austausch Dienst (DAAD) fellowship from June 2005 to December 2006 from Germany to complete part of his doctoral studies at the Physics Department of Technical University of Chemnitz, Germany. He is also the recipient of a number of other national and international fellowships. He received his doctoral degree from Indian Institute of Technology Kharagpur, India in June 2007. His doctoral thesis was selected among the group of top five by the selection committee of Prigogine Prize 2009 towards the choice of best doctoral thesis in Thermodynamics. He taught at Indian School of Mines University Dhanbad, India from March to October 2007. He returned to National Institute of Technology Durgapur, India, again as a permanent faculty in October 2007, and professing till date at the level of Associate Professor in the Department of Mechanical Engineering. He was the panel topper of many faculty selection boards. He has established himself as a perfect pedagogue and fundamental researcher. His primary research interest lies in different areas of Thermodynamics. He conceived the idea of the law of motive force, which is a fundamental law of nature, in 1989 when he was an undergraduate student.

A. K. Pramanick, *The Nature of Motive Force*, Heat and Mass Transfer,
DOI: 10.1007/978-3-642-54471-2, © Springer-Verlag Berlin Heidelberg 2014

The manufacturer's authorised representative in the EU is Springer Nature Customer Service Centre GmbH, Europaplatz 3, 69115 Heidelberg, Germany. If you have any concerns regarding our products, please contact ProductSafety@springernature.com

Printed and bound by CPI Group (UK) Ltd, Croydon, CR0 4YY
19/07/2026
02171041-0001